Advances in Comparative and Environmental Physiology 23

Volumes already published

Vol. 10: Comparative Aspects of Mechanoreceptor Systems
Edited by *F. Ito* (1992)

Vol. 11: Mechanics of Animal Locomotion
Edited by *R. McN. Alexander* (1992)

Vol. 12: Muscle Contraction and Cell Motility:
Molecular and Cellular Aspects
Edited by *H. Sugi* (1992)

Vol. 13: Blood and Tissue Oxygen Carriers
Edited by *Ch.P. Mangum* (1993)

Vol. 14: Interaction of Cell Volume and Cell Function
Edited by *F. Lang* and *D. Häussinger* (1993)

Vol. 15: From the Contents: Salivary Gland Secretion –
Nematocyst Discharge –
Adaptations in Decapodan Crustaceans –
Steroid-Sensitive Areas Mediating Reproductive
Behaviors (1993)

Vol. 16: Ion Transport in Vertebrate Colon
Edited by *W. Clauss* (1993)

Vol. 17: Effects of High Pressure on Biological Systems
Edited by *A.G. Macdonald* (1993)

Vol. 18: Biomechanics of Feeding in Vertebrates
Edited by *V.L. Bels*, *M. Chardon* and *P. Vandewalle* (1994)

Vol. 19: Electrogenic Cl^- Transporters in Biological Membranes
Edited by *G.A. Gerencser* (1994)

Vol. 20: From the Contents: Motile Activities of Fish
Chromatophores – Epithelial Transport of Heavy Metals –
Heavy Metal Cytotoxicity in Marine Organisms –
Comparative Pulmonary Morphology and Morphometry –
Molecular Adaptations in Resistance to Penicillins (1994)

Vol. 21: Mechanisms of Systemic Regulation:
Respiration and Circulation
Edited by *N. Heisler* (1995)

Vol. 22: Mechanisms of Systemic Regulation:
Acid-Base Regulation, Ion Transfer and Metabolism
Edited by *N. Heisler* (1995)

Vol. 23: Invertebrate Immune Responses
Cells and Molecular Products
Edited by *E.L. Cooper* (1996)

Vol. 24: Invertebrate Immune Responses
Cell Activities and the Environment
Edited by *E.L. Cooper* (1996)

Advances in

Comparative and Environmental Physiology 23

Invertebrate Immune Responses

Cells and Molecular Products

Guest Editor: E.L. Cooper

With Contributions by

M. Bilej · E.L. Cooper · K. Hashimoto · M.R. Kanost
Y. Kurosawa · R.L. Raison · P. Roch · T. Sawada
V.J. Smith · S. Tomonaga · L. Tučková · L. Zhao

With 51 Figures

Springer

Guest Editor:

Professor Dr. EDWIN L. COOPER

UCLA Medical Center (CHS)
Department of Neurobiology
10833 Le Conte Avenue
Los Angeles, CA 90095-1763
USA

ISSN 0938-2673
ISBN-13: 978-3-642-79695-1 e-ISBN-13: 978-3-642-79693-7
DOI: 10.1007/978-3-642-79693-7

Softcover reprint of the hardcover 1st edition 1996

Cover design: Springer-Verlag, Design & Productions

Typesetting: Thomson Press (India) Ltd., New Delhi

SPIN: 10120153 31/3137/SPS – 5 4 3 2 1 0 – Printed on acid-free paper

Foreword to the Series

The aim of the series is to provide comprehensive, integrated reviews giving sound, critical and provocative summaries of our present knowledge in environmental and comparative physiology, from the molecular to the organismic level.

Living organisms have evolved a widespread range of basic solutions to cope with the different problems, both organismal and environmental, with which they are faced. A clear understanding of these solutions is of course of fundamental interest for all biologists, zoologically or medically oriented. They can be best comprehended in the framework of the environmental and/or comparative approaches. These approaches demand either wide surveys of animal forms or a knowledge of the specific adaptive features of the species considered. This diversity of requirements, both at the conceptual and technological level, together with the fact that physiology and biochemistry have long been mainly devoted to the service of medicine, can account for the fact that these approaches emerged only slowly amongst the other new, more rapidly growing disciplines of the biological sciences.

The field has now gained the international status it deserves and the organization of a series devoted to it appeared timely to me in view of its actual rapid development and of the interest it arouses for a growing number of biologists, physiologists, and biochemists, independently of their basic, major orientation.

Raymond Gilles

List of Editors

Contents

Cell Products: Natural and Induced as Revealed by Non-specific and Specific Responses Following Antigenic Challenge

Chapter 3

The Prophenoloxidase Activating System:
A Common Defence Pathway for Deuterostomes
and Protostomes?
V.J. Smith

Chapter 4

A Definition of Cytolytic Responses in Invertebrates
P. Roch

Chapter 5

The Immunoglobulin Superfamily:
Where Do Invertebrates Fit In?
Y. Kurosawa and *K. Hashimoto*

Contents for Volume 24

Chapter 3

Lectins: Models of Natural and Induced Molecules in Invertebrates
J.A. Olafsen

Chapter 4

Histocompatibility Reactions in Invertebrates
D.A. Raftos

The Environment: The Consequences of Connections and Missed Signals

Chapter 5

Parasite-Invertebrate Host Immune Interactions
T.P. Yoshino and *G.R. Vasta*

Chapter 6

Environmental Pollution and Toxicity in Invertebrates:
An Earthworm Model for Immunotoxicology
A.J. Goven and *J. Kennedy*

Chapter 7

Invertebrate Neuroendocrine and Immune Systems:
Commonality of Mechanisms and Signal Molecules
C. Franceschi and *E. Ottaviani*

Introduction

E.L. Cooper

The Immunodefense System

Because invertebrates are exceedingly diverse and numerous, estimates reveal nearly 2 million species classified in more than 20 phyla from unicellular organisms up to the complex, multicellular protostomes and deuterostomes. It is not surprising to find less diverse defense/immune responses whose effector mechanisms remain to be completely elucidated. Of course, I am not advocating that the few of us devoted to analyzing invertebrate immunity attempt the Herculean task of examining all these species to uncover some kind of unique response! As these two volumes will reveal, we are doing fairly well in examining in depth only the most miniscule examples of invertebrates, some of which have great effects on human populations such as edible crustaceans or insect pests. This is in striking contrast to the mass of information on the mammalian immune response which has been derived essentially from the mouse, a member of one phylum, Vertebrata, an approach, reductionist to be sure, but one that has served well both the technological and conceptual advances of immunology as a discipline. The essential framework of immunology, the overwhelming burst of results since the 1960s, have emanated primarily from this single animal. We should not forget the thymus and the bird's *bursa of Fabricius*, without which we might have been slower to recognize the bipartite T/B system. Furthermore, without the invertebrates and the prescient observations of Metchnikoff, cells would have played a minor role in immunologic theory or would have been recognized as important much later than at the beginning of this century. Moreover, during the transition from invertebrate phagocytosis to macrophages, the universal innate or natural immune system might have been overlooked and macrophages might also have been late arrivals.

This is the first of two volumes (Vols. 23 and 24 of the series, Advances in Comparative & Environmental Physiology) devoted to the now popular subject of invertebrate immunology. The main focus of Volume 23 is on cells and molecular products. Cells, the basic immunodefense armamentarium, and the antigens to which they respond are examined in Chapters 1 and 2. In the subsequent

Laboratory of Comparative Immunology, Department of Neurobiology, School of Medicine, University of California, Los Angeles, California 90095-1763, USA

Advances in Comparative and Environmental Physiology, Vol. 23

chapters, the immune response is discussed. Cell products are covered in Chapters 3–7. At the end of this Introduction, a selected number of thoughtfully written reviews and books, restricted to the last 5 years, are listed. I take full responsibility for choosing this time span and for not including experimental papers.

What Is This Volume About?

We will skip the protozoans and simple metazoans and leap immediately to the complex metazoans, multicellular forms in which a great division of labor exists, resulting from differentiation and complex organization. Obviously, organisms consist of cells that comprise systems, and the immune system is no exception, as described by Sawada and Tomonaga in Chapter 1. The immunology of invertebrates can progress only as far as the limits of cell and molecular biology allow. First, there is need to create a better climate for analyzing the functional activities of hemocytes by establishing more clearly a unified and acceptable model of the cell, its structure, and its surface markers. To do this, there is need of a concerted international effort to isolate one or two cell types, concentrate on them exclusively, and subject them to a battery of assays that will use, e.g., monoclonal antibodies. Finally, the products of these cells must also be subjected to advanced biochemical and molecular analyses in the attempt to determine the genes responsible for their activities.

The world is replete with foreign material, antigens of various kinds, some of which threaten the life of invertebrates after they have been exposed to them. There are natural antigens, but as Tučková and Bilej (Chap. 2) point out, it has been possible to modify natural antigens by chemical methods to produce molecules that provide a better possibility to distinguish between self and non-self. This undeniable starting point lies at the basis of the immune response, at least for its initiation. Inherent in this absolute requirement is the need for recognition and the assurance that foreign material, antigens, will bind to an appropriate receptor on a potential effector cell or that there is the equivalent of an antigen-processing cell that will capture foreign antigens, rendering them palatable to an effector cell. This would serve to set the immune response in motion. At this point we can give some consideration to different forms of immunodefense responses.

How the proPO system evolved depends upon information concerning enzyme activity in living species and how the phylogenetic relationships among all the invertebrates are interpreted. Clearly, proPO-type immunodefense as measured by phenoloxidase activity is a rather universal strategy that has been often studied in the arthropods, notably in insects and crustaceans, as presented by Smith (Chap. 3). With regard to the less studied species, however, the presence of prophenoloxidase and associated protease is found only among the ascidians. As with other mechanisms and from a comparative point of view, much remains to be learned concerning the proPO system when representative protostome and deuterostome species are compared. This unique system reinforces the constant dilemma faced by invertebrate immunologists, i.e., to distinguish between responses that are directly related to host defense (innate, natural, nonadaptive) and those that are immunologic (induced, adaptive). In the future, we will need more thorough biochemical analyses that define molecular mass, subunit structure, and amino acid sequence. Moreover, knowing these answers should then lead to an analysis of nucleotide sequence and ultimate isolation, as well as to the cloning of the genes that control the proPO system.

The lytic system most closely resembles a condition in vertebrates by way of the complement pathway, and its relevance relates to patterns that are found among vertebrates, particularly in fishes and amphibians (see the thorough analysis by Roch, Chap. 4). Moreover, there is strong evidence for the resemblance of invertebrate hemolysin and vertebrate pore-forming proteins. At this point, we can accept a putative functional homology as fairly clear or as having evolved separately. However, the necessary structural and coding information that dictates and regulates the lytic process must be clearly defined. At best, if we are to assume strong linkages, the most acceptable hypothesis for now seems that the perforins of invertebrates are a more acceptable alternative to choosing a likeness (homology?) with complement. However, it must be borne in mind that biochemical and molecular analyses are essential if we are to draw clear distinctions between mechanisms of hemolytic-cytolytic activity in invertebrates, a diverse and complicated group.

Clearly, the Ig superfamily is a group of molecules that were first defined in vertebrates as proteins, so classified because they possess the Ig fold, as discussed by Kurosawa and Hashimoto (Chap. 5). These molecules are involved in recognition events, especially those involving protein-protein recognition. Although first defined as crucial to the immune system, interesting and fairly complete analyses reveal an apparent similarity between members of the immune system and those that act primarily during recognition in the nervous system, where they evolved as early as the Cnidaria. Thus, there are points in evolution where there is a similarity between recognition molecules that lead to developmental pathways, and those that may be shunted toward the evolution of immunodefense. The fine line that distinguishes between the shunting in either or both of these directions is a challenge for those interested in the Ig superfamily (see also Chapter 7).

A partial clarification of this dual dilemma facing proponents of the Ig superfamily is revealed by an analysis of insect hemolymph proteins from the insect Ig superfamily. Consideration is given to proteins that are involved in the insect's immune apparatus, such as hemolin. However, there are still many open questions regarding the outcome of those members of the Ig superfamily that are involved in developmental events as they relate to the development of the nervous system vs those regarding the functions that are primarily of an immunodefensive nature. Here, one is left to ponder the concomitant development of the three great regulatory systems: the nervous, immune, and endocrine systems that serve to balance the internal milieu. (See also Chapter 7 in Volume 24 for a broad coverage of this intraorganismic network.)

Cell-free immunity in insects includes inducible antimicrobial peptides that are the chief means of defense against invading organisms (see Chap. 6 by Kanost and Zhao). There is ample evidence characterizing the molecular biology and genetics of immune peptides available in species such as *Cecropia*, *Sarcophaga*, *Manduca*, and *Drosophila*. The cecropins and other inducible, antibacterial proteins are believed to act as effector molecules of the immune system that may be partially analogous to the perforins and the complement system in vertebrates. These insect defensins show significant homologies to mammalian (including human) microbial peptides present in polymorphonuclear leukocytes and macrophages. Although defensins have been discovered only recently, the presence of homologues in certain invertebrates suggests that they are ancestral components of the host defense system and the complement system in vertebrates.

This brings us to a critical interface where points of union between the invertebrates and vertebrates come together and distinctions are blurred. The question of complement and its evolution is a crucial one directed at the immunodefense apparatus. It is therefore appropriate that we look carefully at what complement is like in a vertebrate that bridges the gap, so to speak, between advanced vertebrate forms and invertebrates. Since hagfish represent a reasonably well-studied group of vertebrates that are clearly the most primitive, it is somewhat consoling that the problem of complement has been in part resolved (see Chap. 7 by Raison). The solution has not come from what is decidedly invertebrate as the controversy surrounding perforin vs complement is revealed by invertebrate lytic mechanisms. Rather, this solution comes from having convincingly clarified what was once considered an IgM molecule in the hagfish; but which is, in fact, a complement component. Moreover, the relationship of the hagfish serum protein to higher vertebrate humoral factors has been elucidated almost simultaneously by two groups, using molecular approaches. Similarity was shown when the hagfish protein was compared to the mammalian complement components C3, C4, and C5, using PCR, mRNA and nucleotide sequences. The broad functional activity of these components of humoral immunity is related to that of opsonins, known to be one of the prime functions of humoral components among invertebrates. In essence, immune functions are often classified as innate, nonadaptive mechanisms. It would be exciting to link these two instances of opsonic activity, i.e., that of the invertebrate components with that of the primitive component as revealed by hagfish, to produce a unifying system of molecules that are involved in lytic activity and other functions assigned to humoral components of invertebrates.

Perspectives

All vertebrates and invertebrates manifest self/nonself recognition. Any attempt to answer the question of the adaptive significance of recognition must take into account the universality of receptor-mediated responses. This may take two forms: (1) rearrangement of clonally distributed antigen-specific receptors that distinguish in the broadest sense between self and nonself, and between nonself A and nonself B, latecomers on the evolutionary scene; (2) pattern recognition receptors, the earliest to evolve and still in existence, necessitating induced second signals in T- and B-cell activation. Neither strategy forces the organization, structure, and adaptive functions of vertebrate immune systems upon invertebrates. Thus, we can freely delve into the unique aspects of the primitive immune mechanisms of invertebrates. In contrast, using the opposite strategy, which is still problematic, i.e., linking invertebrate and vertebrate defense, seems to give us an approach to universality that might eventually reveal homologous kinship. For those interested further in the problem of immunologic theory as it relates to invertebrates, the following references will be valuable.

References

List of selected pertinent references published within the last 5 years pertaining to these two volumes

Beck G, Cooper EL, Habicht GS, Marchalonis JJ (1994) Primordial immunity, foundations for the vertebrate immune system. New York Academy of Sciences, New York, 376 pp

Cooper EL (1992) Overview of immunoevolution. Boll Zool 59: 119–128

Cooper EL (1994) Invertebrates can tell us something about senescence. Aging Clin Exp Res 6: 3–23

Cooper EL, Nisbet-Brown E (1993) Developmental immunology. Oxford University Press, New York, 480 pp

Cooper EL, Rinkevich B, Uhlenbruck G, Valembois P (1992) Invertebrate immunity: another viewpoint. Scand J Immunol 35: 247–266

Ganz T, Lehrer RI (1994) Defensins. Curr Opinions Immunol 6: 584–589

Garside P, Mowat A McI (1995) Polarization of Th-cell responses: a phylogenetic consequence of nonspecific immune defense. Immunol Today 16: 220–225

Hoffman JA (1995) Innate immunity of insects. Curr Opinions Immunol 7: 4–16

Hoffmann JA, Janeway C, Natori S (1994) Perspectives in immunity: the insect host defense. RG Landes Co, Austin

Hultmark D (1993) Immune reactions in *Drosophila* and other insects: a model for innate immunity. Trends Genet 9: 178–183

Humphreys T, Reinherz EL (1994) Invertebrate immune recognition, natural immunity and the evolution of positive selection. Immunol Today 15: 316–320

Janeway CA Jr (1992) The immune system evolved to discriminate infectious non-self from non-infectious self. Immunol Today 13: 11–16

Marchalonis JJ, Schluter SF (1990) On the relevance of invertebrate recognition and defense mechanisms to the emergence of the immune response of vertebrates. Scan J Immunol 32: 13–20

Quintans J (1994) Immunity and inflammation: the cosmic view. Immunol Cell Biol 72: 262–266

Smith LC, Davidson EH (1992) The echinoid immune system and the phylogenetic occurrence of immune mechanisms in deuterostomes. Immunol Today 13: 356–362

Stewart J (1992) Immunoglobulins did not arise in evolution to fight infection. Immunol Today 13: 396–399

Tauber AI (1994) The immune self: theory or metaphor? Immunol Today 15: 134–136

Vetvicka V, Sima P, Cooper EL, Bilej M, Roch P (1993) Immunology of annelids. CRC Press, Boca Raton, 300 pp

Cells: The Basic Immunodefense Armentarium

Chapter 1

The Immunocytes of Protostomes and Deuterostomes as Revealed by LM, EM and Other Methods

T. Sawada[1] and *S. Tomonaga*[2]

Contents

[1] Department of Anatomy, Yamaguchi University School of Medicine, 1144 Kogushi, Ube, Yamaguchi 755, Japan
[2] School of Allied Health Science, Yamaguchi University, 1144 Kogushi, Ube, Yamaguchi 755, Japan

Advances in Comparative and Environmental Physiology, Vol. 23

1 Introduction

The immunocytes of invertebrates have not been well studied, since only a limited number of species of arthropods, mollusks, echinoderms and urochordates have been investigated. Most species of invertebrates do not have blood vessels that isolate the circulating fluid and cells from other connective tissue components. Thus, wandering cells in the fluid circulating through vessels or filling spaces among tissue cells and the coelomic cavity are presented in this chapter.

Hemocytes, the major cellular components in the immune system, were first studied by light microscopy, and their fine structure was later studied by electron microscopy (excellent reviews in Ratcliffe and Rowley 1981a, b; Cohen and Sigel 1982; Ratcliffe et al. 1985). These reviewers found that basic observations on morphology seemed to have been accomplished, but that invertebrate immunology still involved much confusion and controversy, especially with respect to hemocyte classification or identification. Not only do we need more morphological information, but there is also a dearth of data concerning hemocyte behavior and function. Characteristics revealed by molecular biology and ontogeny will be required to further clear up this confusion and controversy. In this chapter, we present a summary of the present situation of invertebrate immunology mainly with regard to hemocytes.

2 Immune Mechanism and Immunocytes

Here, immunocytes are taken to include all cells which mainly function in immunodefense mechanisms. Thus, not only the cells involved in adaptive immune responses but also those performing phagocytosis or secreting abtibacterial substances are included. However, the boundaries between the immunodefense system and other physiological systems have become unclear owing to the recent expansion of information that indicates a close relationship between the immune system and neural or endocrine systems (Blalock 1989; Teschemacher et al. 1990). Immunodefense mechanisms in invertebrates also show close relations with the neuroendocrine system (Stefano et al. 1989, 1991; Hughes et al. 1992). We will review here mainly those cells involved in phagocytosis, cytotoxicity and secretion of defense molecules upon contact with foreign organisms. The representative cell groups that play immunodefensive roles are free wandering cells existing in hemolymph, blood, coelomic cavity and connective tissue. Those cells contained in hemolymph are usually called "hemocytes" or "blood cells" in invertebrates. They vary somewhat from phylum to phylum or species to species, regarding their functions, behavior and biochemical features. However, many of them participate in the immunodefense mechanism by various means: phagocytosis, encapsulation, cytotoxicity and secretion of antibacterial substances (Ratcliffe et al. 1985).

2.1 Hemocytes and Coelomic Cells

In spite of the numerous species of invertebrates only a relatively small number of studies have been made on invertebrate hemocytes and immune systems in comparison to what has been achieved in even one species of mammal. In addition, studies on invertebrate hemocytes tend to be distributed unequally. The information from those studies has been available on a limited number of species, a limited subgroup of hemocytes and a limited variety of immunological aspects. Thus, we have constructed invertebrate immunology on a rather restricted number of species and a limited amount of evidence. Consequently, it is probable that a lack of basic information causes misunderstandings, confusion or controversies in analyzing invertebrate immune responses.

Although it is not yet possible to describe all aspects or general features of invertebrate hemocytes, the efforts of invertebrate immunologists have brought about convincing progress in establishing a basis for the immunology of invertebrate hemocytes. We try to summarize the work on invertebrate hemocytes in spite of the controversy and confusion which are still present even in hemocyte the classification and nomenclature. We do not describe studies of hemocytes in detail here, because there are already excellent works reviewing such studies in most animal phyla (such as reviews edited by Ratcliffe and Rowley 1981a, b; Cohen and Sigel 1982). Basic morphology seems to have been well studied already in most animal phyla, but using only representative species. First, we try to outline the hemocyte types in some invertebrate phyla. However, because each phylum involves considerable variation from species to species, descriptions of size and detailed morphology are not applicable to every species in a particular phylum.

2.1.1 Wandering Cells in Noncoelomic Animals

Complex metazoans without coelomic cavities (the acoelomates) have interstitial cells in mesoglea or mesodermal parenchyma, between ectoderm and endoderm. Interstitial cells are not hemocytes but are potential candidates as ancestral cells of hemocytes in more advanced animals. In sponges these cells are phagocytic, and some of them may play a role in glycogen storage as glycocytes (van de Vyver 1981). Wandering cells are relatively rare in the coelenterates and may be absent in the Hydrozoa. Interstitial cells in the mesoglea are sometimes self-proliferating and some species also have glycocytes (van Praet and Doumenc 1974; van de

Vyver 1981). Flatworms possess wandering cells, ameboid and phagocytic, in mesodermal parenchyma, but wandering cells of nematodes do not seem to be phagocytic and their functions are unknown (van de Vyver 1981).

2.1.2 Blood Cells (Hemocytes) of Protostomes

2.1.2.1 Annelids

The hemocytes of annelids have been classified roughly into two groups (Cooper and Stein 1981; Dales and Dixon 1981): ameboid cells which possess pseudopodia and are potentially phagocytic (amebocytes), and large spherical cells which are characterized as nutritive reservoirs especially of lipid droplets (eleocytes in Polycheta and Oligocheta, chloragogen cells in Hirudineans). Amebocytes may include several cell types with respect to size and granular inclusions, but the relationships between these several different types remain obscure. There is speculation that different types of amebocyte may represent several intermediate stages in the course of differentiation of a single cell lineage (Dales and Dixon 1981).

Eleocytes or chloragogen cells are large (15–70 μm) and characterized by an accumulation of lipid, carotenoid pigment and glycogen (Dales 1961; Eckelbarger 1976). These cells are sometimes referred to as granular cells but they differ morphologically from granular amebocytes. In addition, there is speculation that eleocytes in some species may be derived from phagocytic amebocytes (Dales and Dixon 1981). Cooper and Stein (1981; Stein et al. 1977; Stein and Cooper 1978) attempted to consider cytochemical aspects regarding the classification, and divided hemocytes into further subgroups: basophils, neutrophils, acidophils, granulocytes, and chloragogen cells.

Cells described above as hemocytes are found in both coelomic fluid and blood of annelids, suggesting that hemocytes migrate frequently into and out of blood vessels. A few annelids also have erythrocytes which contain proteins for oxygen transport or storage (Terwilliger et al. 1985).

2.1.2.2 Mollusks

Gastropods. Hemocytes of gastropods may be divided into four groups; one is a group of relatively small, round cells with a high nucleus-to-cytoplasm ratio, and the other three are groups of spreading cells with pseudopodia and high variability in morphology (Sminia 1981). Small round cells are considered as young cells capable of proliferating (Brown and Brown 1965). The last three are classified into granulocytes, hyalinocytes and amebocytes. Granulocytes are large cells, up to 70 μm in diameter; they extend pseudopodia (Yoshino 1976), and contain electron-dense granules which are sometimes positive for acid phosphatase (Cheng and Garrabrant 1977). Hyalinocytes and amebocytes are similar in size, i.e., about 8 μm in diameter (Yoshino 1976). Both are small spherical cells that extend pseudopodia after spreading on glass slides, but amebocytes may spread to a length of 70 μm (Sminia 1981). Granulocytes and amebocytes contain a Golgi complex and lysosomes that are positive for acid phosphatase and peroxidase. Amebocytes may contain two populations, spreading and round that differ in behavior and cytochemistry. Hyalinocytes differ from granulocytes and amebocytes, and form a separate cell group (Sminia 1981).

Bivalves and Cephalopods. Bivalves have two types of hemocytes: hyalinocytes and granulocytes, each of which may be further divided into three groups depending on species. Hyalinocytes are divided into types I, II and III, and granulocytes into types A, B and C (Cheng 1981). Cephalopods have been reported to possess a single type of hemocyte which contains many granular components and Golgi complexes (Cowden and Curtis 1981).

2.1.2.3 Arthropods

Arthropods have an open circulation system and numerous hemocytes are contained in their hemolymph. Six hemocyte types are the main constituents of insects according to Jones (1962), and the Jones' system has been also adopted for noninsect arthropods such as onychophorans and myriapods. Hemocytes of crustaceans, another representative group of arthropods, however, are composed of fewer groups, about three. Insects and crustaceans constitute two major groups in arthropods. In addition, the insects are one of the best investigated groups of invertebrates whose hemocytes have been investigated extensively with respect to immune competence.

Insects. Prohemocytes, plasmatocytes, granular cells, cystocytes, spherule cells and oenocytoids are the cells that compose the insect hemocyte population (Jones 1962; Price and Ratcliffe 1974; Rowley and Ratcliffe 1981). Prohemocytes are spherical cells with a high nuclear-to-cytoplasm ratio and have been considered as undifferentiated and proliferative cells, with a diameter of 6–13 μm in the wax moth *Galleria mellonella* (Price and Ratcliffe 1974). Plasmatocytes, usually spindle-shaped with a 10–15 μm diameter, spread on glass slides to become ameboid cells and possess lysosomal enzymes in several species (Rowley and Ratcliffe 1981). Granular cells, containing many cytoplasmic granules of different electron density, are spherical cells of 8–20 μm diameter. They rapidly degranulate (Rowley and Ratcliffe 1976; Rowley 1977) and extend many filopodia during larval stages (Wago 1982). Cystocytes, round cells of 8–15 μm diameter, contain many granules that are fragile and easily exocytosed as they are removed from insect bodies (Rowley and Ratcliffe 1981). Spherule cells are of a diameter of 8–16 μm and contain 1–3 μm spherules, easily identifiable by a characteristic feature as they contain crystal-like structures (Rizki 1962). Oenocytoids, up to 30 μm, contain many rod-like granules (Akai and Sato 1973; Lai-Fook 1973).

Crustaceans. Three main groups of hemocytes have been found in crustaceans (Bauchau 1981): hyaline cells, semigranular cells and granulocytes. Semigranular cells are the group which exhibits intermediate size and an abundance of cytoplasmic granules between the other two types. Hyaline cells are the smallest hemocytes and contain few organelles. Granulocytes are the largest cells whose cytoplasm is filled with large granules and scattered free ribosomes. Thus, these three hemocyte groups seem to show continuous differentiation within the granulocyte series (Cuénot 1891; Wood and Visentin 1967; Bauchau and De Brouwer 1972; Bodammer 1978), but this speculation needs further experimental analysis. Identification and distinction of hemocytes using newer staining methods have also been attempted in order to classify hemocytes by the cytochemical characteristics of their cytoplasmic components (Hose et al. 1987). Moreover, recent functional approaches led to a conclusion that hyaline cells are involved in the initiation of hemolymph coagulation while granulocytes are involved in defense against foreign material by phagocytosis and encapsulation (Hose et al. 1990).

2.1.3 Blood Cells (Hemocytes) of Deuterostomes

2.1.3.1 Echinoderms

Phagocytic amebocytes, spherule cells, vibratile cells and progenitor cells are probably the most common cell types found in the coelomic fluid, and they are usually considered as hemocytes in echinoderms (Smith 1981). Phagocytic amebocytes are large cells, with a 14–30 µm diameter (Bookhout and Greenburg 1940; Bertheussen and Seljelid 1978), and are the most common cells found in all species. They exhibit two different forms: petaloid and filopodial (Chien et al. 1970; Fontaine and Lambert 1977; Edds 1977a, b, 1980; Otto et al. 1979). In vitro, they are petaloid in hanging-drop cultures (Johnson 1969a) and are filopodial on glass slides (Bertheussen and Seljelid 1978). Spherule cells (8–20 µm in diameter) are filled with numerous, large spherical inclusions that may contain mucopolysaccharides and protein (Hetzel 1963; Fontaine and Lambert 1977). Vibratile cells seem to be spherical structures with a flagellum (Hetzel 1963; Bertheussen and Seljelid 1978) and have cytoplasmic granules that may contain polysaccharides (Johnson 1969b; Vethamany and Fung 1971). Progenitor cells are small and spherical (6–8 µm in diameter) containing a large nucleus (with prominent nucleoli) which is surrounded by a thin, hyaline cytoplasm (Hetzel 1963; Fontaine and Lambert 1977); they are considered to be coelomocyte stem cells (Endean 1958; Hetzel 1965; Ratcliffe and Rowley 1979). Two other coelomocytes have been reported in some restricted echinoderm groups: hemocytes (10–23 µm) containing hemoglobin (Hetzel 1963; Fontaine and Lambert 1973), and crystal cells (20–24 µm) that include rhomboidal or star-shaped crystals (Hetzel 1963; Johnson and Beeson 1966; Fontaine and Lambert 1977).

2.1.3.2 Urochordates

Hemocytes of tunicates may be classified roughly into five groups (Wright 1981). Hemoblasts are small spherical cells with relatively high nucleus-to-cytoplasm ratio and are believed to be hemopoietic stem cells (Ermak 1976; Milanesi and Burighel 1978). Cells that contain lysosomes or small vesicles and a scarce supply of large granules are usually referred to as hyaline cells or hyaline amebocytes and tend to flatten and spread extensively on glass slides (Fuke 1979; Sawada et al. 1991, 1993; Zhang et al. 1992; Fuke and Fukumoto 1993). Hyaline cells, in some species, may be considered as precursors of granular cells (Kalk 1963). Granular cells are filled with numerous granules that sometimes occupy most of the cell volume. They possess a centrally located nucleus (Overton 1966; Sawada et al. 1991; Zhang et al. 1992; Fuke and Fukumoto 1993). Vacuolated cells possess one or several large vacuoles that occupy most of the cell volume, with subgroups designated according to the classical terminology: signet-ring cells with a single large vacuole and an eccentrically located nucleus, compartment cells and morula cells. The last two subgroups have multiple vacuoles with no apparent, distinct definition between them, neither functionally nor cytochemically (Wright 1981). In most species, active phagocytes are hyaline and/or granular cells. Cytochemical studies of granular cells and vacuolated cells have been unable to reveal their functions.

In many species, each of the hemocyte groups mentioned above includes further multiple groupings, resulting in a total number of up to ten hemocyte types. Certain species have been reported to have nephrocytes that contain calculus bodies inside large vacuoles (George 1930, 1939; Ohuye 1936; Andrew 1961; Vallee 1967), but the function of nephrocytes and the differences between them and other vacuolated cells are not clear.

3 Cells Participating in Immune Reactions

Commonly known immunodefensive reactions of invertebrates consist of phagocytosis, cytotoxicity, encapsulation, and secretion of material to damage foreign organisms or to modulate immunocyte functions. In addition, certain immune activities may involve other systems as well. For example, hemocyte activity has been proved to be modulated by various substances associated with the neuroendocrine system (Stefano et al. 1989, 1991; Hughes et al. 1992; Cooper et al. 1993) Therefore, neuroendocrine cells in this sense may also be referred to as components of the immunodefense system. This expansion of immunological information strongly suggests that the word "immunocyte" may contain certain ambiguities. Nevertheless, we prefer to follow certain more classically accepted criteria in this chapter, leaving the secretion of cytokines or neuroendocrine substances in immunocyte functions to other chapters.

3.1 Hemocytes in Immune Responses

3.1.1 Phagocytic Cells

Because it is a universal and well-known phenomenon, phagocytic activity of immunocytes against vertebrate red blood cells, yeast particles, carbon particles, latex beads and bacteria has been examined in many invertebrates (Ratcliffe et al. 1985). Various cells of invertebrates are clearly phagocytic without showing any precise relation with any abundance of granular components of the cytoplasm. In particular, fine granules such as carbon particles appeared to be ingested by most ameboid and granular cells. However, phagocytosis involving large particles such as erythrocytes or yeast particles seems more prominent in the cells with hyaline cytoplasm than in granular cells, as seen in the tunicates *Halocynthia roretzi* (Sawada et al. 1991; Zhang et al. 1992; Fuke and Fukumoto 1993) and *Styela clava* (Sawada et al. 1993). Active phagocytes have a tendency to exhibit significant motility due to ameboid movement or to spread into thin cytoplasmic sheets on glass slides.

There are close relations between phagocytosis and humoral factors in hemolymph. Many invertebrates contain opsonins which accelerate the incorporation of foreign particles, as reported in a protochordate by Kelly et al. (1993). Migration of phagocytes would also be affected or controlled by humoral factors. In the land slug *Incilaria fruhstorferi*, injection of latex beads or yeast particles into the hemocoel induced an increase in phagocytic cells contained in the fluid of the hemocoel (Furuta and Shimozawa 1987, 1994). This increase is a result of active migration of surface-lining cells of hemocoel and fibroblast-like cells from the surrounding tissue (Fig. 1). Such an explosive increase in migrating activity of phagocytic cells into the hemocoel strongly suggested the presence of cytokine-like humoral control on phagocyte activity in those animals.

3.1.2 Cytotoxic Cells

Invertebrate cells are capable of killing other cells in a process known as cytotoxicity (Cooper 1992). Cytotoxicity against mammalian erythrocytes has been reported in some invertebrates (Wittke and Renwrants 1984), particular cell lines (Tyson and Jenkin 1974; Decker et al. 1981) and allogeneic cells (Bertheussen 1979; Kelly et al. 1992). Effector cells in such cytotoxicity were different types depending on the case and species, i.e., granular cells (Janeway 1989; Porchet-Hennere et al. 1992), sipunculid leukocytes (Boiledieu andValembois 1977a, b; Valembois and Boiledieu 1980), or phagocytes (Bertheussen 1979). With respect to the evolutionary origins of cytotoxic cells, such cells seem to have few common morphological characteristics. On the other hand, multiple hemocyte types or even most hemocytes exhibited cytotoxicity against allogeneic or xenogeneic cells in several cases (Tyson and Jenkin 1974; Fuke 1980). Furthermore, evidence for defining particular cytotoxic cell types (i.e., effector cells) was often insufficient

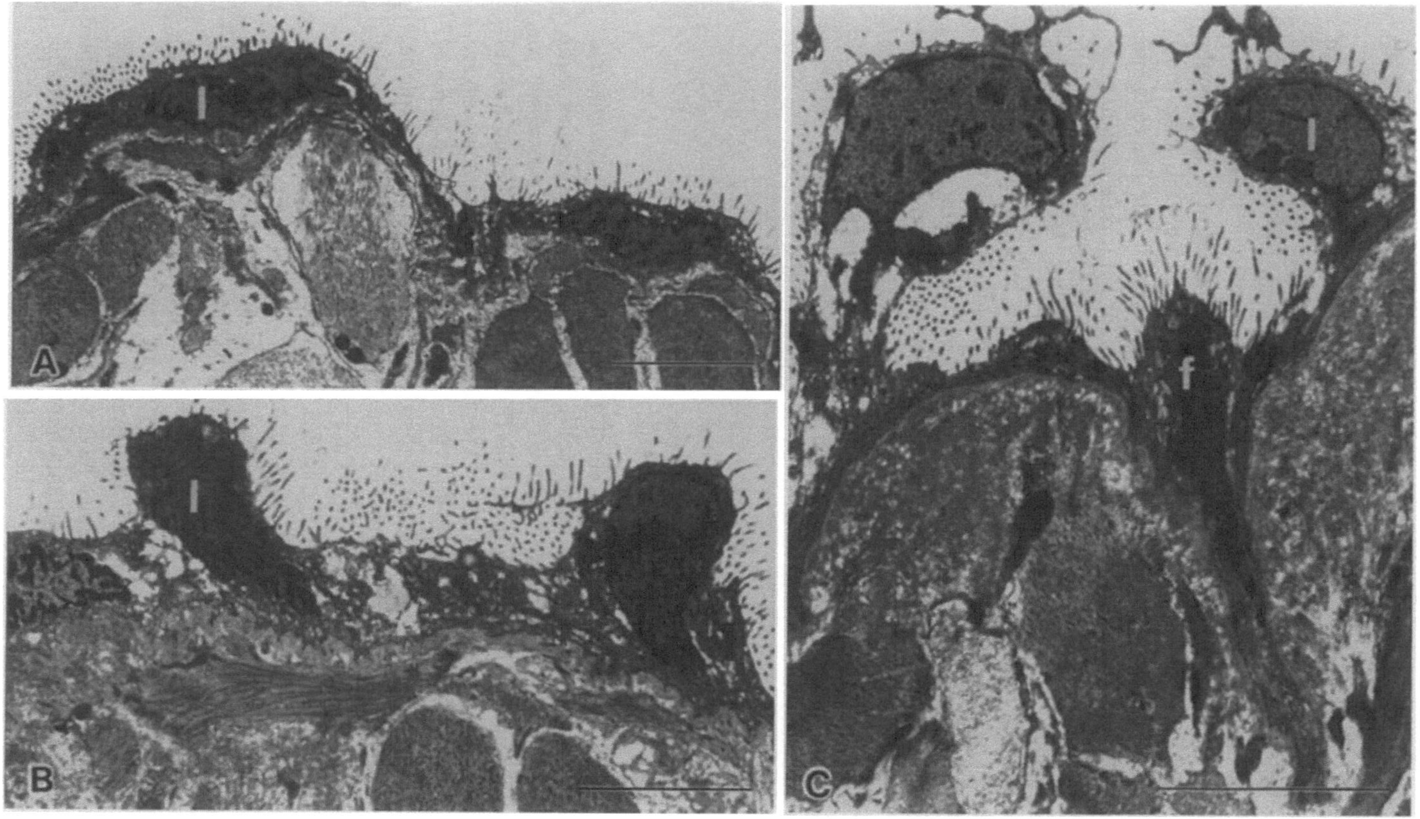

Fig. 1. Migration of the lining cells of coelomic cavity and of the fibroblastic cells after injection of yeast particles into hemocoel of the land slug *Incilaria fruhstorferi* (Gastropoda). **A** Lining cells (*l*) in normal land slug were flat cells with microvilli, having cytoplasm of relatively high electron density. The tissue surrounding the cavity was constituted of muscle and connective tissue. **B** The lining cells (*l*) protruded into the cavity 1 h after injection. **C** 3 h after injection, lining cells (*l*) left the wall of the cavity and fibroblast-like (*f*) cells migrated out of surrounding tissue. *Bars* = 5 µm. (Courtesy of Dr. E. Furuta)

(Parinello et al. 1993). Detailed processes and molecular mechanisms of cytotoxicity also remain mostly unknown.

Recently, studies of cytotoxic phenomena among invertebrates have been concerned with the evolutionary development of vertebrate natural killer cells (Janeway 1989; Evans and Cooper 1990; Franceschi et al. 1991; Cooper 1992). However, the cells which correspond to mammalian cytotoxic lymphocytes or natural killer cells have not been proven to exist in invertebrates. In previous studies of many invertebrate species, small spherical hemocytes were most frequently designated as hemoblasts, lymphocytes or lymphocyte-like cells and have been considered as the most likely candidates for hemopoietic stem cells. However, in tunicates, small spherical cells different from hemoblasts at least with respect to nuclear components may exist in some species. There are reports indicating proliferation after allogeneic stimuli (Raftos and Cooper 1991) and infiltration of lymphocyte-like cells into first and second allografts (Raftos et al. 1987). These small spherical cells of tunicates may include certain cytotoxic cell populations, other than those designated as hemopoietic stem cells (Ermak 1976; Sawada et al. 1993). In the annelid *Nereis diversicolor* granular cells exhibited natural killer activity using a pore-forming protein (Porchet-Hennere et al. 1992). Thus, it is still problematic whether the morphological similarity between invertebrate cytotoxic cells and vertebrate NK-cells is valid with respect to their functions.

3.1.3 Exocytotic Granular (Vacuolated) Cells

For cytotoxicity to occur, for instance, effector cells must discharge cytoplasmic components to effect a "hit" to target cells. The discharge of granules from vacuoles by hemocytes was also observed when hemocytes were exposed to endotoxin, foreign materials or foreign cells (Fuke 1980; Armstrong and Rickles 1982). Such granular (vacuolated) hemocytes filled with granules (vacuoles) are commonly found in many invertebrates (Rowley and Ratcliffe 1981). Some of them as viewed by microscopy discharge their granules at once as if to explode, while others tend to excrete them one by one. Certain granular cells are highly active in ameboid movement, and others are completely inactive.

The discharge of cytoplasmic granules (degranulation) is not itself an immunodefense function. Whether it is so naturally depends upon the components contained in the cytoplasmic granules (vacuoles). The contents of cytoplasmic granules have been investigated cytochemically with regard to lysosomal enzymes. Various kinds of substances including antibiotics, antimicroorganismal substances, enzymes and cytotoxic factors have been detected in invertebrate granular hemocytes. However, in most invertebrates, factors contained in the cytoplasmic granules have not yet been characterized except for a few examples in which the contents have been investigated exceptionally well. For example, granulocytes of the horseshoe crab contain the factors for coagulation, precursors of various enzymes, and antibacterial molecules

(Mürer et al. 1975; Nakamura et al. 1985; Muta et al. 1990; Toh et al. 1991). In insect plasmatocytes and oenocytoids (Ashida et al. 1988), prophenoloxidase has been detected.

Difficulties in purification of particular types of hemocytes, especially the spontaneous degranulation of granular cells during the procedure, often allow only a minimal biochemical investigation of granule contents in many species. So far, only a few invertebrate granulocytes have been characterized with respect to their granular contents. Therefore, there have been persistent arguments about many questions regarding invertebrate granular cells: the functions and roles in the immunodefense mechanisms of each species, the phylogenic evolution of granular cells, as well as the relationships between vertebrate and invertebrate granular cells are points that rest upon meager evidence and are still open to much speculation.

3.1.4 Cells Involved in Nodule Formation or Encapsulation

Nodule formation and encapsulation are also representative immunodefensive functions in invertebrates (Ratcliffe et al. 1985). The immunocytes involved in this process are the hemocytes (or coelomocytes) which adhere to each other or on to the surfaces of foreign particles. They are elongated chloragocytes and round leukocytes in annelids (Valembois et al. 1992). In insects, granular cells, plasmatocytes and cystocytes are involved in encapsulation (Sato et al. 1976; Schmit and Ratcliffe 1978; Ratcliffe et al. 1985). In mollusks, the responsible cells are amebocytes (Sminia et al. 1974; Cheng and Garrabrant 1977). Hemocytes of echinoderms (Johnson 1969c) and urochordates (Wright and Cooper 1975; Parinello et al. 1977) also encapsulate.

In urochordates, many studies have been accomplished by inserting foreign particles into the tunic (tissue outside the ectoderm). We recently observed the formation of nodule-like structures by a single hemocyte type in *Halocynthia roretzi* and *Styela clava* during in vitro culture in plastic plates without adding any foreign particles. However, in comparison, hemocytes inside this structure showed a close similarity to those involved in encapsulation or nodule formation, forming multiple layers of elongated cells which form tight contacts with each other (Sawada et al., in prep.).

3.1.5 Alloreactive Cells

The ability to recognize allogeneic differences may not be unique to immunocytes. Allogeneic differences are obviously recognized by oocytes and spermatozoa of certain hermaphroditic and self-sterile animals during fertilization (Morgan 1938; Fuke 1990). Moreover, not all immunocytes may possess the ability to recognize allogeneic cells. Mammalian macrophages or granulocytes have been believed to be unable to recognize allogeneic markers, although recent studies proposed the presence of an alloreactive macrophage in mice (Yoshida et al. 1991). Also, in invertebrates, it may be reasonable to suppose the occurrence of certain hemocytes which do not recognize allogeneic targets by themselves, despite the insufficient information presently available on the alloreactivity of all of the hemocyte types. On the other hand, allogeneic recognition actually seemed an important and characteristic function of cells involved in immunodefense mechanisms (Ey and Jenkin 1982; Ratcliffe et al. 1985; Cooper 1982). Despite many studies of allograft rejection in invertebrates (Cooper 1968, 1969; Hildemann et al. 1980; Ratcliffe et al. 1985; Kelly et al. 1992), cytological and molecular evidence is insufficient in many cases to identify conclusively the particular cells responsible for rejection.

Here, an essential question arises about why this function is required. Invertebrate hemocytes in vivo would seldom encounter allogeneic cells under normal conditions, except for a few situations such as fusion between tunicate (Urochordate) colonies. In most cases, typical phenomena representing allogeneic recognition occurred only under experimental conditions such as allograft rejection.

3.2 Hemopoiesis

Because invertebrate hemocytes are essential components of immunocytes, hemopoiesis is important in invertebrate immunodefense. So far, hemopoiesis in invertebrates does not tend to be restricted to specific hemopoietic organs. No rule seems to exist proposing that more evolved species have a specific hemopoietic organ more frequently. However, there are several exceptional examples for invertebrate hemopoietic organs.

Situated behind the eyes of cephalopods (squid and octopus) are the well-known hemopoietic organs designated as white bodies (Cowden and Curtis 1974). The axial organ in starfish (echinoderms) has been considered as an immune organ that contains cells performing specific immune responses and possessing characteristics similar to vertebrate lymphatic organs (Leclerc et al. 1986). Rather well-developed hemopoietic tissues superior to the stomach (or epigastric region), surrounding the lateral arterial vessel, and also within the maxilliped have been reported in penaeid shrimp (Crustacea; Arthropoda) (Bell and Lightner 1988). Kondo et al. (unpubl.) also confirmed the presence of certain multiple hemopoietic sites during ontogenetic development in the Kuruma prawn, *Penaeus japonicus* (Fig. 2). It has been reported that the ridgeback prawn (*Sicyonia ingenitis*) has a unique hemopoietic organ (a pair of hemopoietic nodules) adjoining the anteroventral portion of the hepatopancreas (Martin et al. 1987). This particular organ is commonly present in penaeid shrimp and it is generally called the lymphoid organ (Oka 1969; Bell and Lightner 1988) although it contains no lymphoid cells. To the best of our knowledge the so-called lymphoid organ in penaeid shrimp is likely to play an important role in antigen-trapping mechanisms rather than serving as a hemopoietic tissue (Kondo et al. 1994).

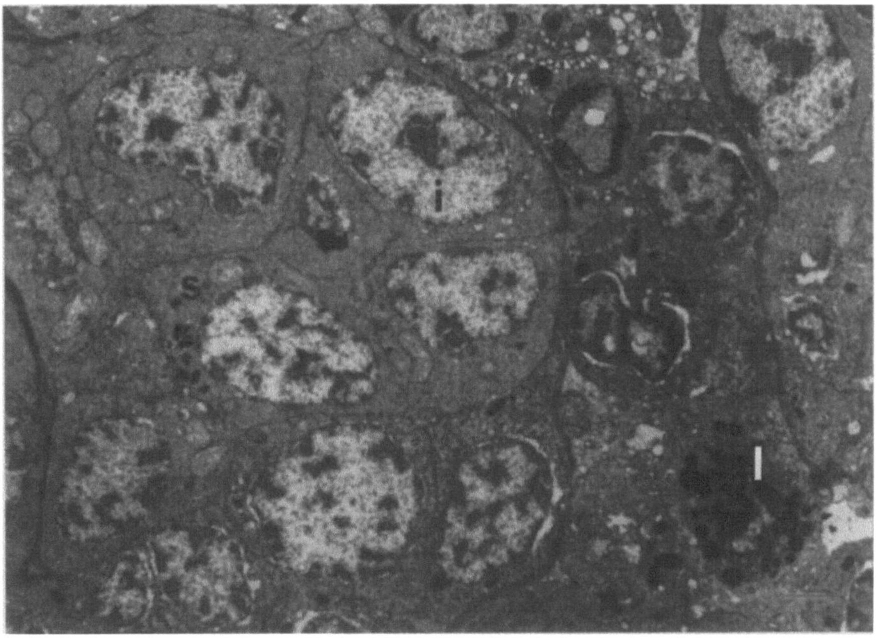

Fig. 2. An electron micrograph of epigastric hemopoietic tissue of kuruma prawn, *Peneus japonicus*. *i* Immature hemocytes; *l* large granular cell; *s* small granular cell. *Bar* = 1 μm

Except for several invertebrates mentioned above, hemopoietic tissues have not been identified in many invertebrates. This may support the view that many of invertebrates without apparent specific hemopoietic organs may rely on interstitial connective tissue, including hemolymph as a field of hemopoiesis. So far, we have little knowledge concerning microenvironments for invertebrate hemopoiesis such as the stroma and the extracellular matrix, including humoral factors supporting hemocyte growth and differentiation. These unknown factors may restrict the site for hemocyte proliferation and maturation, and there may also be fewer restrictions within the interstitial tissues at sites of hemopoiesis that we have so far not detected.

3.3 Nonhemocyte Cells in Defensive Functions

Cells in the hemolymph, the circulating fluid or the fluid filling in tissue spaces, together with free cells found among connective tissues, are referred to as hemocytes. These hemocytes may be mainly responsible for immunodefense mechanisms. However, cells other than hemocytes or undifferentiated interstitial cells of lower invertebrates seem to accomplish some defensive functions which involve the process of self and nonself recognition. We would like to describe two unique nonhemocyte cells which may have certain relationships to hemocytes in their function and ontogenic development.

3.3.1 Cnidocytes of Sea Anemones

It is not legitimate to consider coelenterate cnidocytes as immunocytes using the same criteria as for hemocytes. Cnidocytes are cells that apparently do not wander nor circulate within the hemolymph. However, cnidocytes originate from interstitial, undifferentiated cells (Lehn 1951; David and Murphy 1977), and discharging nematocysts seem to resemble degranulating hemocytes. In addition, such nematocysts are effective defense tools against potential threats, which are equally important as their functions as a mechanism for catching prey. In several species at least, nematocysts have been shown to participate in attack against allogeneic individuals by releasing their contents (Watson and Mariscal 1983a, b). Clearly, though, this defense system is not useful against bacterial infection or internal parasites.

The sea anemone *Haliplanella luciae* discharges its stenotele lining on its feeding tentacles not only when the tentacles contact prey but also when they contact allogeneic tentacles. This attack by nematocysts is followed by a further behavioral reaction that allows either of the allogeneic individuals to escape. Then, each animal develops the so-called catch tentacles several days after the allogeneic contact. These catch tentacles are derived from regular, feeding tentacles, now bearing new and larger nematocysts specialized for attacking allogeneic individuals. These nematocysts are insensitive to prey animals and are released only on contact to allogeneic animals (Watson and Mariscal 1983a, b). This allospecific response includes allogeneic recognition on the surface of both feeding and catch tentacles, and includes a change in the nematocyst-producing program after the first encounter with allogeneic animals. Thus, the phenomena appear similar to what happens in the cells of the immune system and common problems may underlie this superficial similarity.

3.3.2 Tunic Cells of Tunicates

Tunicates (Urochordata) have a tunic, a thick layer composed of mucopolysaccharides, outside their ectoderm. In the tunic, there are many and varied types of cells (Endean 1961; Deck et al. 1966; Wardrop 1970; De Leo et al. 1981; Hirose et al. 1991), which apparently have a role in protecting the body underlying the tunic. Certain tunic cells often share morphological similarities with hemocytes and at least some of the tunic cells have been considered as hemocytes in origin (Wright 1981).

Tunic cells are responsible for the formation of tunic (Endean 1961; Zaniolo and Trentin 1987) which protects soft tissues from mechanical stress. Second, they participate in the immunodefense system against invasion by microorganisms, since the tunic could favor the growth of microorganisms. The tunic must contain certain cells which patrol and clear potentially pathogenic organisms. Consequently, it is a place where the "inflammatory-like response" against foreign material occurs, as revealed in the results of experiments (Parinello et al. 1984). In another situation, the tunic cells are involved in allogeneic discrimination at colony fusion of colonial species (Taneda et al. 1985; Hirose et al. 1988), which is a rather unique example of cells encountering allogeneic cells under natural conditions.

4 Phylogenic Lineages of Immunocytes

One of the interesting subjects of comparative immunology is the question of how immunocytes of different phyla are related to each other and how vertebrate immunocytes have developed in phylogeny (Klein 1989; Cooper et al. 1992a). Hemocytes are probably a major component in immune reactions, and they have been studied in each animal phylum, as briefly summarized in this chapter. Within each animal phylum hemocytes may be classified into several subgroups with more or less common nomenclature (Rowley and Ratcliffe 1981), in spite of certain controversies concerning detailed classification schemes. However, there is no well-established and common category of hemocyte nomenclature nor classification scheme over different phyla. The schemes used for hemocyte classification in protostomes seem to resemble each other more than those used for deuterostomes. In deuterostomes, echinoderms may have four subgroups of hemocytes, whereas urochordates exhibit so much variation in hemocyte composition that hemocyte types can hardly be equated even between two different species. In contrast to urochordates, various vertebrates seem to have a more uniform hemocyte composition. The variety of the urochordate life styles (solitary and colonial, asexual reproduction in various processes, etc.) may have required such a great variety in hemocytes. In spite of little evidence for interspecies relationships, it may still be acceptable to suppose that certain phylogenic lineages of immunocytes took part in invertebrate evolution. Such lineages may be very complicated, and such a simple process as dividing a single stem into multiple branches can hardly be expected. Advanced species often have a simpler hemocyte composition than lower species.

Morphological similarity between two different hemocyte subgroups, such as occurs in the granular contents of the cytoplasm, is not sufficient evidence to categorize them as related groups. Functional characteristics, including classical ones such as ameboid movement and nodule formation, and information at a molecular level are the aspects that take priority in arguments about hemocyte classification and evolutional relationships. Still we can offer some speculation when we consider evolutionary concepts such as homology/analogy (Cooper 1992; Cooper et al. 1992b).

4.1 Cell Functions as Markers for Phylogenic Lineages

Functional similarity is an important aspect when phylogenic relationship of homocytes are being considered, but some similarities are too universal to be specific markers of phylogenic lineage.

Phagocytic hemocytes are probably present in most invertebrates. Most hemocyte types are more or less phagocytic in certain invertebrates. In addition, a peculiar localization of fixed phagocytic cells in the wall of arterioles was reported in many species of decapod crustaceans (Cuénot 1905; Johnson 1987). Thus, phagocytic activity is a kind of universal function among various invertebrate hemocytes, so that this function alone cannot be a specific marker for phylogenic lineage. Speculation may only be possible if combined with some morphological aspect, e.g., that hyaline cells (hyalinocytes) with cytoplasm having few large granules but containing lysosome-like vesicles tend to be most active in phagocytosis. These hyaline cells, especially of lower deuterostomes, are the most probable candidates for a phylogenic lineage to which vertebrate macrophages might be related.

Granulated (vacuolated) cells are also present in most invertebrates, regardless of the enormous variations in cell size, number and size of cytoplasmic granules (vacuoles), and behavior. Theoretically, certain lineages of granulated and vacuolated hemocytes should exist in phylogeny. Nevertheless, enormous variations among granulated (vacuolated) cells have prevented us from correlating them even within a single phylum, family or order, as shown in tunicates (Wright 1981). Consequently, morphological differences and similarities should not be overestimated when consideration is given to criteria upon which to base phylogenic lineage. Arguments about their relationships in phylogeny, beyond the levels of phyla, will become fruitful at least after the molecular characteristics of the contents of cytoplasmic granules (vacuoles) are revealed. So far, unfortunately, such information has been insufficient except for a few invertebrates.

Nodule formation and encapsulation may be used as functions on which certain hemocyte lineages could be based. These functions seem to be more evident in protostomes, annelids, arthropods and mollusks, although deuterostomes also exhibit similar functions. However, the cells involved in encapsulation or nodule formation appear to include two different types phagocytes and cells of a fibroblastic nature. In addition, the occurrence of phenoloxidase in the blood and prophenoloxidase in the hemocytes of the ascidian *Ciona intestinalis* (Jackson et al. 1993) may provide a clue for the similarity of cellular responses in encapsulation and nodule formation between protostomes and deuterostomes.

4.2 Ontogeny

Ontogeny is another important aspect in the argument regarding phylogenic hemocyte or immunocyte lineages. Differentiation pathways in particular should be studied more in order to answer questions concerning immunocyte evolution, because only little is known about invertebrate hemopoiesis. What we can speculate about at this point may be that cells referred to as "immunocytes" in invertebrates and vertebrates are of mesodermal or mesenchymal origin (Du Pasquier 1992). In ontogeny, cells of mesoderm or mesenchyme exhibit a greater capacity to behave independently and to migrate away from cell masses to which they originally belonged. This behavioral character of mesodermal or mesenchymal cells fits the demand for immunocytes, since cells of the immunodefense system should migrate rapidly into an infected region or should patrol throughout the body creating a mobile surveillance system. Therefore, a general origin of immunocytes may be traced back to mesenchymal cells in sponges, primitive multicellular animals, or even to the unicellular ameba.

5 Various Approaches to Immunocytes

In contrast to previous investigators, modern immunologists have applied electron microscopy, flow cytometry, monoclonal antibody analysis, cell culture, etc. to analyze invertebrate immunocytes from different viewpoints. In this section, we introduce some of these approaches. However, even after the fine structure of hemocytes has been reported in most experimental animals, light microscope observations remain of basic importance, constituting the easiest and most important method of examining cells, especially living cells, in experiments. We describe first an example of how regular light microscope observations on living cells associated with short-term cultures have recently been applied and have served as the basis for more advanced studies.

5.1 Light Microscope Studies on Tunicate Hemocytes

Hemocytes of a tunicate *Halocynthia roretzi* were first described briefly by Ohuye (1936) and then in more detail by Fuke (1980). The descriptions and classification schemes of these anthors were acceptable to a certain extent, but were still insufficient as a base for further experimental studies of hemocyte function and behavior. In fact, these classifications resulted in some confusion, particularly under experimental conditions using viable hemocytes, because they lacked information about morphological and behavioral variation of hemocytes. Such information is required for precise identification of hemocyte types in experiments since viable hemocytes modify their shape significantly in vitro. We therefore needed to establish a method as well as a criterion for identifying living hemocytes after continuous observation in short-term cultures.

Recently, several schemes of hemocyte classification were proposed (Sawada et al. 1991; Azumi et al. 1993; Fuke and Fukumoto 1993). Two of them (Sawada

Table 1. Correspondence between several classification schemes of *Halocynthia roretzi*

Light microscopy Sawada et al. (1991)	Electron microscopy Fuke and Fukumoto (1993)	Zhang et al. (1992)	Azumi et al. (1993)	Ohtake et al. (1989)
p1 cell[a]	Hyaline amebocyte?	Phagocyte	Type A?	Small granular amebocyte
p2 cell	Hyaline amebocyte	Phagocyte	Type C?	Small granular amebocyte
g1 cell	Granular amebocyte	Granulocyte small granule	Type B?	Large granular amebocyte
g2 cell	Large basophilic cell	Basophilic cell		
g3 cell	Large granular cell	Fibrous material-containing cell		
v1 cell	Vacuolated cell type 1	Vacuolated cell type 1?		
v2 cell	Macrogranular cell globular cell	Vacuolated cell type 2		
v3 cell	Vacuolated cell type 2	Vacuolated cell type 1	Type F	
v4 cell	Vacuolated cell type 3	Vacuolated cell type 2?	Type F	
ly cell	Lymphocyte like cell	Lymphoid cell	Type D?	

[a] p1, p2 cell: phagocyte types 1, 2; g1, g2, g3 cell: granular cell types 1, 2, 3; v1, v2, v3 cell: vacuolated cell types 1, 2, 3; ly cell: lymphocyte-like cell.

et al. 1991; Fuke and Fukumoto 1993) included information about morphological modifications of living hemocytes in short-term culture. Two additional studies (Ohtake et al. 1989; Zhang et al. 1992) were electron microscope studies, and another one by Azumi et al. (1993) was an attempt at separating several hemocyte groups by density gradient and equating the biomolecular nature of hemocyte subpopulations to their electron microscope images. Naturally, these schemes should be correlated with each other. The correlation between different schemes would become perfect only after hemocyte functions are totally understood. However, the investigators seemed to have advanced enough to correlate their classification schemes as far as the morphological aspects are concerned (Table 1), except that cell sizes in Sawada et al. (1991) should be reduced.

In this process, light microscope studies on viable hemocytes by regular, phase-contrast, Nomarski interference and fluorescence microscopy, especially in combination with short-term culture and video-recording techniques, were the most effective ways of obtaining important information for finding correspondences between different images of the same hemocytes. Many of the images after shape changes in vitro could be seen to be those of a single hemocyte only by continuous observation. At the same time, many were distinguishable only by their behavior. For example, three types of granular cells (g1, g2, g3 cells; Sawada

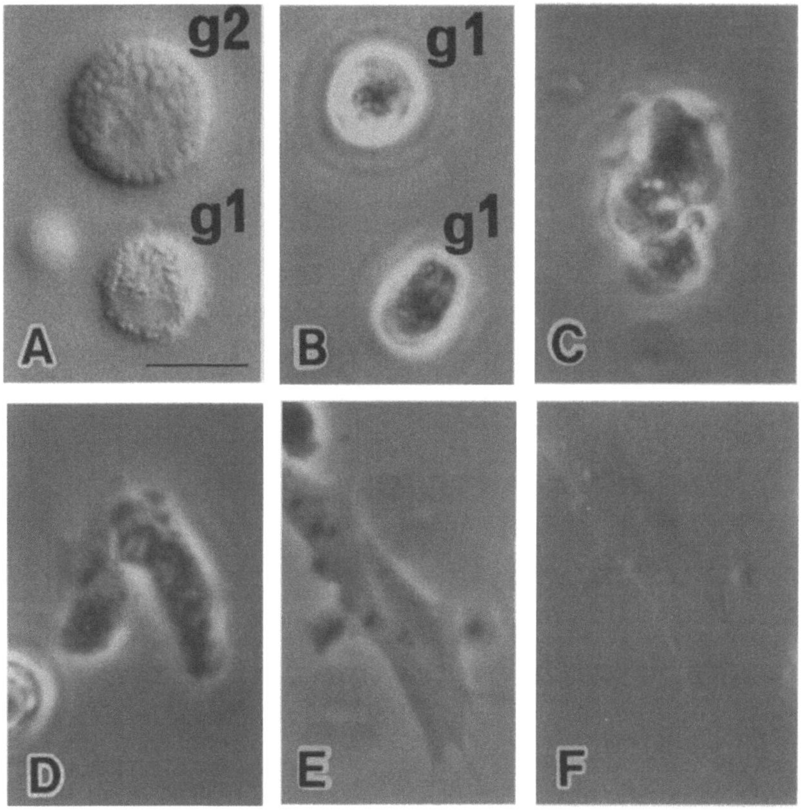

Fig. 3. Granular cell type 1 (g1 cell) of *Halocynthia roretzi*. **A** Granular cells type 1 (*g1*) and type 2 (*g2*) were filled with cytoplasmic granules, and differed in size (Nomarsky interference microscopy). **B–E** (phase contrast): g1 cells were round at first (**B**), tended to aggregate together, and spread (**C, D**) finally into rod shapes with many spine-like protrusions (**E**). **F** Nomarsky interference microscopy of the same cells as in **E**. Note that g1 cells lost their cytoplasmic granules during spreading. *Bar* = 10 μm

et al. 1991) seemed to differ only in their sizes unless they extended on glass slides (Figs. 3, 4, 5). Among them, g3 cells were acidophilic but both g1 and g2 cells were basophilic in Wright's or Giemsa staining. Slight differences in the size of cytoplasmic granules did not seem sufficient to separate g1 and g2 cells (Fig. 3). However, the morphology and the behavior of g1 and g2 cells appeared completely different after they extended on glass slides (Figs. 3, 4). G1 cells tended to aggregate together, spread in a rod-like shape with spine-shaped cytoplasmic extensions, and exhibited poor mobility. In contrast, g2 cells spread as extraordinary large sheets (in fact, this is the largest of all hemocyte types after it has spread) in a fan or crescent shape, exhibiting active movement. The behavior of g3 cells (Fig. 5), in vitro was also unique as they are very quickly transformed into a worm-like shape, and they never spread into flat sheets.

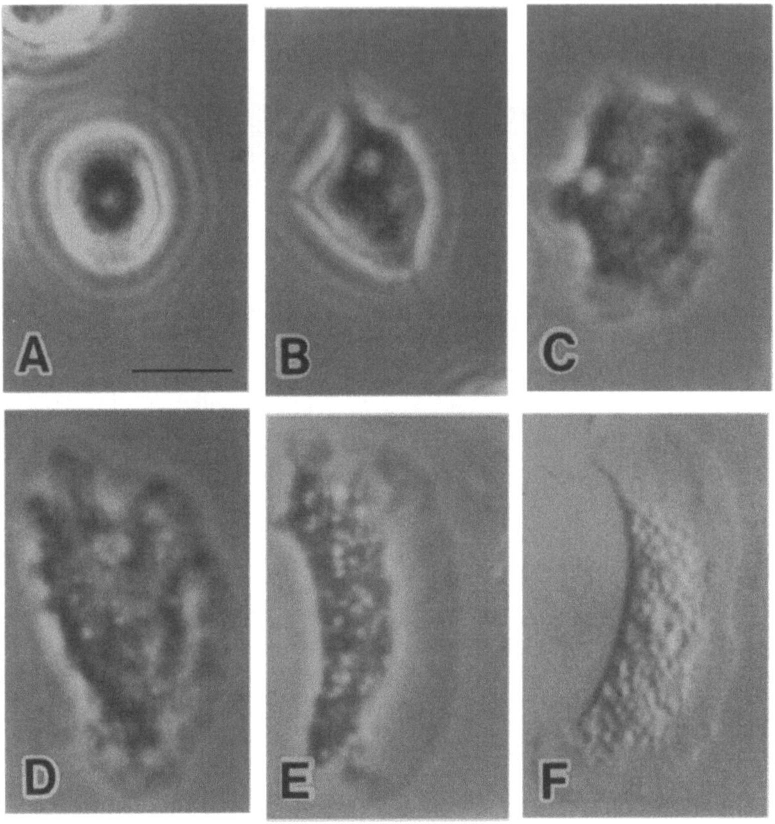

Fig. 4. Granular cell type 2 (g2 cell) of *Halocynthia roretzi* spreading on a glass slide. **A–E** (phase contrast) Round cells with dark contrast (**A**) started spreading (**B**,**C**,**D**) and finally became a thin and fan-shaped cytoplasmic sheet (**E**). **F** Nomarsky interference microscopy of a spread of g2 cells showed that it retained many cytoplasmic granules after spreading. *Bar* = 10 μm

Although several hemocyte types of *Halocynthia roretzi* still remain uncorrelated between the different classification schemes, an effort to investigate the morphological variety of the same hemocyte type under various conditions, may be important as a foundation for further experimental studies. Reexamination of preexisting classification schemes has also been attempted in *Styela clava* (Sawada et al. 1993) and *Ascidia sydneiensis samea* (Michibata et al. 1990) based on recent results dealing with hematopoiesis (Sawada et al., in prep.) and vanadium concentration in hemocytes (Michibata et al. 1990), respectively. In many invertebrate species, reviewing previous descriptions, reconstitution of hemocyte classification schemes and establishment of cell identification systems suitable for viable cells are necessary in order to promote future studies on hemocyte functions and differentiation both in vitro and in vivo.

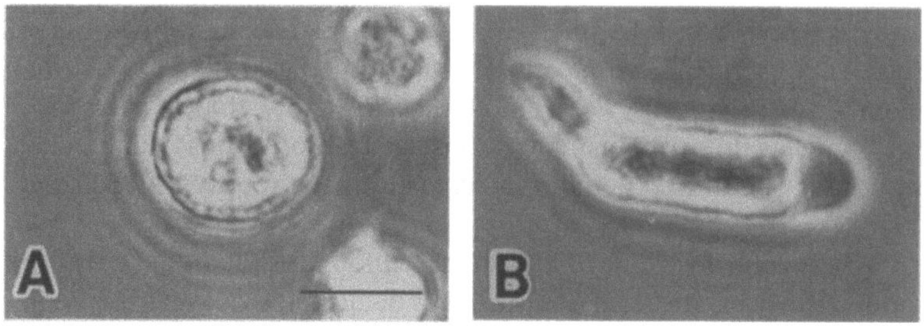

Fig. 5. Granular cell type 3 (g3 cell) of *Halocynthia roretzi* (phase contrast). **A** Unless g3 cell started to move, it was round and highly refractive. **B** g3 cells moved fast and in a worm-like shape unique to them. *Bar* = 10 μm

5.2 Fluorescence Microscopy

Fluorescence is very useful for identifying living cells once we have efficient markers to distinguish cell types. Specific cell groups may be marked with fluorescein-labeled monoclonal antibodies, lectins, or other fluorescent dyes. Marked cells can be distinguished under fluorescence microscopy or can be analyzed by a fluorescence-activated cell sorter (FACS). Certain fluorescent dyes are incorporated by viable cells and become useful markers. Moreover, autonomous fluorescence has sometimes been an effective marker for distinguishing hemocyte types (Deno 1987; Michibata et al. 1988, 1990). When such autonomous fluorescence corresponds to a particular cell type, it will be the most useful marker of the cell type during continuous observation under viable conditions. Thus, we can avoid the risk of modifying cell functions induced by binding antibodies or lectins on the cell surface.

Recent progress in the technique of measuring chemiluminescence involving little energy will make the chemiluminescence assay a useful method for revealing hemocyte functions. This has been already proven in the investigations on the oxygen-radical generating system in phagocytes of the snail *Lymnaea stagnalis* (Adema et al. 1991) and about hydrogen peroxide production by phagocytes of the sea urchin *Strongylocentrotus nudus* (Ito et al. 1992). This technique in combination with computer image analysis may open a new field devoted to the morphological analysis of invertebrate immunocytes.

5.3 In vitro Culture

The culture of cells, tissues and organs will solve many questions dealing with hemopoiesis, cell differentiation, cell classification and cell functions. Cultures of insect hemocytes are the best-studied examples in invertebrates (Ratcliffe et al. 1985) and several analyses have provided information on hemocyte cultures including the composition of the culture medium (Landureau 1968; Landureau and Grellet 1975). Furuta (1989) and Furuta and Shimozawa (1983) reported successful culture of connective tissue from the land slug (Mollusk) in which macrophage-like cells were maintained for more than 3 months. The culture medium was unique in using trehalose instead of glucose. In annelids, mitogenic responses of cultured hemocytes have also been reported (Roch and Valembois 1978).

Until recently, in vitro culture of hemocytes was not successful in urochordates. However, successful culturing of hemopoietic tissues has been reported in the tunicate *Styela clava*, confirming that hemocyte proliferation occurred even after 3–4 weeks inside cultured pharynx tissue (Raftos et al. 1990; Sawada et al. 1994). In contrast, most hemocytes usually degenerate and disappear during 1–2 weeks of culture, when hemocytes were cultured alone (outside pharynx tissue) in vitro even in the same medium as was successful for tissue culture (Raftos et al. 1990). In short-term incubation, for 12–24 h, hemocytes certainly exhibited proliferative activity as measured by uptake of ^{3}H-thymidine (Sawada, unpubl. data). Thus, apparently hemopoietic cells require additional supplies of growth factors from tissue cells in order to maintain proliferation for long durations. To this end, IL-1-like molecules have been proved to be present in tunicates (Raftos et al. 1991) as in other invertebrates (Beck and Habicht 1986, 1991). Tunicate IL-1 is probably one of those hemopoietic growth factors.

Continuous proliferation of stem cells accompanied by hemocyte differentiation, together with colony formation of mature hemocytes, will be an ultimate goal of hemocyte culture and of many attempts to control hemocyte differentiation under artificial conditions. However, more information about growth factors, cellular adhesion, and extracellular matrix molecules will be required to accomplish this final purpose.

5.4 Monoclonal Antibodies

Monoclonal antibodies, which have been used as very effective tools for the identification of mammalian hemocytes, have been less fruitful in invertebrate immunology so far. There are too few biologists working on each species to produce and examine various potentially useful monoclonal antibodies. Therefore, monoclonal antibodies have not yet become standard in identifying invertebrate immunocyte types. Nevertheless, we can describe certain successful studies using monoclonal antibodies, and the number of other analyses using them is certainly increasing. In *Lymnaea stagnalis* (Gastropoda), monoclonal antibodies have been utilized to identify hemocyte subpopulations in juvenile and adult animals (Dikkeboom et al. 1988). Subpopulations of hemocytes have also been distinguished in *Biomphalaria glabrata* (Gastropoda) by revealing cell-surface antigens to which monoclonal antibodies were specific (Yoshino and Granath 1983, 1985).

A word of caution, however, is essential because certain monoclonal antibodies possibly cross-react widely over many antigenic molecules, since these antibodies recognize only a single epitope composed of several amino acids or monosaccharides. For instance, a monoclonal antibody UB-15, which we made, recognizes fibroblast-like cells (granular cells type 1) among the hemocytes of *Halocynthia roretzi*, and it cross-reacts not only with other tunicate hemocytes but also with cells of vertebrate tissues including mammalian thymic epithelial cells. However, the relationship between UB-15 antigens in vertebrates and in tunicates is unknown so far. The staining pattern may suggest that UB-15 antigens are associated with cytoskeletons whereas tunicate plasma also contains the components that bear the UB-15 antigenic epitope (Sawada, unpubl. data). Thus, precise information and identification of antigenic epitopes may not be possible in some cases when monoclonal antibodies are actually applied to experimental studies. Investigators must carefully restrict the situation, procedure and purpose for using monoclonal antibodies in order to avoid confusion in their studies caused by the cross-reactivity.

5.5 Flow Cytometry

Flow cytometry has provided a breakthrough for analyzing extremely large number of cells, since 10–20 000 cells can be examined in several minutes. However, this technique is not as often used in the immunology of invertebrates as in mammals, and even standard profiles with forward- and right-angled light scattering of hemocytes have not been established in most species. Even simple analysis of cell size (measured by forward-angled light scattering) and surface nature (measured by right-angled light scattering) of hemocytes would be useful for detecting statistical differences in hemocyte composition under experimental conditions. For instance, the forward- and right-angled light scattering profile of several fractions of tunicate hemocytes, separated in a bovine serum albumin (BSA) density gradient, certainly indicated that each fraction was constituted of different cell populations (Azumi et al. 1993), However, more experiments are essential to properly scrutinize the data since morphological changes in a response to BSA must be considered. Nevertheless, these results are still enough to demonstrate the effectiveness of flow-cytometry analysis.

The first technical problem that investigators must resolve to apply flow-cytometry analysis to invertebrate hemocytes is coagulation. Hemocytes or hemolymph of most invertebrates coagulate easily when they are handled in vitro. An anticoagulant is inevitably present if a calcium-free medium is being used, as is the case in many studies. Azumi et al. (1993) actually added EDTA to the medium as anticoagulant. However, we mention here the possibility that an anticoagulant such as a calcium-free medium sometimes caused certain morphological changes in the hemocytes; fine and spine-like filopodia may become bleb-like protrusions, for example. Our data showed a time-dependent shift of the profiles of forward- and right-angled light scattering analysis, indicating that hemocytes of *Halocynthia roretzi* changed their morphological characteristics in the medium containing the anticoagulant EGTA (Fig. 6).

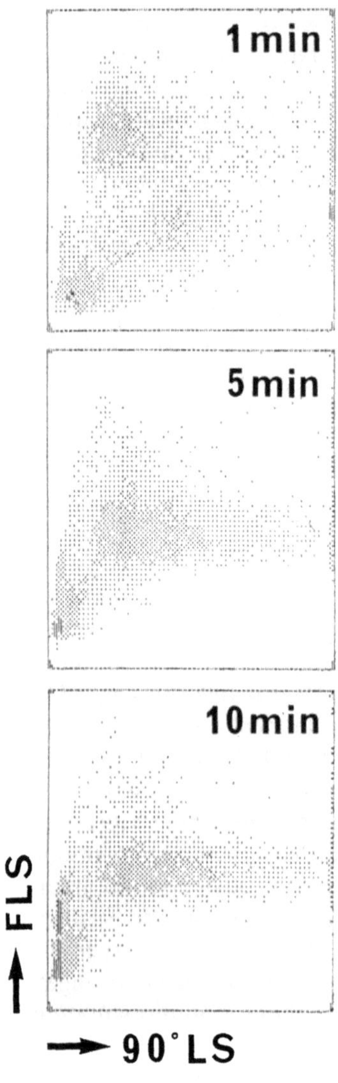

Fig. 6. Change in flow-cytometry profiles in forward- and right-angled light scattering (*FLS* and *90°LS*) analysis. Hemocytes were analyzed 1 min, 5 min and 10 min after they had been harvested into anticoagulant (EGTA) solution. Three groups could be distinguished in the profile at 1 min, but only two groups were apparent in the profile at 10 min

Appropriate labeling of cells, by immunofluorescence using monoclonal antibodies for example, would multiply the repertoire of experiments to which this method can be applied. However, it is also true that flow cytometry alone cannot reveal any significant findings, unless the data are effectively combined with those derived from other techniques. The successful development of flow cytometry in studies of mammalian lympho-hematopoietic cells has been supported by the development of stable and well-characterized monoclonal antibodies as cell differentiation markers. In invertebrates, there are several examples of the unique and useful application of flow cytometry to hemocytes. Flow cytometry in combination with phagocytotic analysis using FITC-labeled particles led to the successful detection of a stimulative effect of fluid factors on phagocytic activity of earthworm hemocytes (Bilej et al. 1990).

5.6 Electron Microscopy

Regular observation by transmission electron microscopy (TEM) appears to be completed in most invertebrate phyla, although controversies remain. To clear these problems, TEM must be combined with scanning electron microscopy (SEM), cell purification techniques and immunohistochemical staining methods. EM technique, as well as light microscopy and its modifications, will continue to be a fundamental and effective tool in experimental studies.

6 Problems and Possible Solutions

Research in invertebrate immunology is progressing in diverse fields. However, there are still many new questions being raised as well as fundamental problems left unsolved. First, there is still confusion and even controversy regarding hemocyte classification in many species. Those problems will be solved only after various types of information are accumulated from experiments using a single species, by investigators working on different aspects of hemocyte function. Second, as a natural consequence, separation and purification of hemocyte populations have not been performed so that biochemical characterization and mass analyses of functional problems are presently still difficult. Third, stable markers specific for hemocyte types, other than morphological characteristics, are inadequate to perform functional studies since hemocytes change their shape and cytoplasmic contents considerably. One possible solution here is the widespread development and extensive use of monoclonal antibodies. Fourth, information about the microenvironment surrounding hemocytes and supporting their functions, such as the extracellular matrix and cell-regulating factors (e.g. cytokines), is sparse in invertebrates. However, recent progress in analyzing invertebrate fibronectins, cytokines, coagulation factors and the prophenyloxidase cascade will certainly provide the molecular basis for more sophisticated studies of cell functions.

Finally, studies concerning hemocyte ontogeny are related to all four of the above problems. At present, although easy to imagine, it is still difficult to clarify the hemopoietic cell lineages and differentiation stages of hemocytes. However, in future, that hemopoiesis and ontogenic lineage of hemocytes may be studied with more intensity in relation to the phylogeny or evolution of immunocytes. This would create, in our opinion, one of the most interesting subjects in comparative immunology, particularly in the four problem areas of invertebrate immunology just given above.

References

Adema CM, van Deutekom-Mulder EC, van der Knaap WPW, Meuleman EA, Sminia T (1991) Generation of oxygen radicals in hemocytes of the snail *Lymnaea stagnalis* in relation to the rate of phagocytosis. Dev Comp Immunol 15: 17–26

Akai H, Sato S (1973) Ultrastructure of the larval hemocytes of the silkworm, *Bombyx mori L.* (Lepidoptera: Bombycidae). Int J Insect Morphol Embryol 2: 207–231

Amen RI, Tijnagel JMG, van der Knaap WPW, Meuleman EA, de Lange-de Klerk ESM, Sminia T (1991) Effects of *Trichobilharzia ocellata* on hemocytes of *Lymnaea stagnalis*. Dev Comp Immunol 15: 105–115

Andrew W (1961) Phase microscope studies of living blood cells of the tunicates under normal and experimental conditions, with a description of a new type of motile cell appendage. Q J Microsc Sci 102: 89–105

Armstrong PB, Rickles FR (1982) Endotoxin-induced degranulation of the *Limulus* amebocyte. Exp Cell Res 140: 15–24

Ashida M, Ochiai M, Niki T (1988) Immunolocalization of prophenoloxidase among hemocytes of the silkworm, *Bombyx mori*. Tissue Cell 20: 599–610

Azumi K, Satoh N, Yokosawa H (1993) Functional and structural characterization of hemocytes of the solitary ascidian *Halocynthia roretzi*. J Exp Zool 265: 309–316

Bauchau AG (1981) Crustaceans. In: Ratcliffe NA, Rowley AF (eds) Invertebrate blood cells, vol 2. Academic Press, London, pp 385–420

Bauchau AG, de Brouwer MB (1972) Ultrastructure des hemocytes d'*Eriocheir sinensis*, Crustace Decapode Brachoure. J Microsc (Paris) 15: 171–180

Beck G, Habicht GS (1986) Isolation and characterization of a primitive interleukin-1-like protein from an invertebrate, *Asterias forbesi*. Proc Natl Acad Sci USA 83: 7429–7433

Beck G, Habicht GS (1991) Purification and biochemical characterization of an invertebrate interleukin 1. Mol Immunol 28: 577–584

Bell TA, Lightner DV (1988) A handbook of normal penaeid shrimp histology. World Aquaculture Society. Allen Press, Lawrence, Kansas, 114 pp

Bertheussen K (1979) The cytotoxic reaction in allogeneic mixtures of echinoid phagocytes. Exp Cell Res 120: 373–381

Bertheussen K, Seljelid R (1978) Echinoid phagocytes in vitro. Exp Cell Res 111: 401–412

Bilej M, Scherlinck J-P, VandenDriessche T, de Baetselier P, Vetvicka V (1990) The flow cytometric analysis of in vitro phagocytic activity of earthworm coelomocytes (*Eisenia foetida*; Annelida). Cell Biol Int Rep 14: 831–837

Blalock JE (1989) A molecular basis for bidirectional communication between the immune and neuroendocrine systems. Physiol Rev 69: 1–32

Bodammer JE (1978) Cytological observations on the blood and hemopoietic tissue in the crab *Callinectes sapidus*. I. The fine structure of hemocytes from intermolt animals. Cell Tissue Res 187: 79–96

Boiledieu D, Valembois P (1977a) Natural cytotoxic activity of spipunculid leukocytes on allogenic and xenogin erythrocytes. Dev Comp Immunol 1: 207–216

Boiledieu D, Valembois P (1977b) The mechanism of leukocyte cytotoxicity studied by time-lapse microcinematography and its inhibition: an example of in vitro specific recognition in invertebrates. In: Solomon JB, Horton JD (eds) Developmental immunology. Elsevier, Amsterdam, pp 51–57

Bookhout CG, Greenburg NP (1940) Cell types and clotting reactions in the echinoid, *Mellita quinquiesperforata*. Biol Bull 79: 309–320

Brown AC, Brown RJ (1965) The fate of thorium dioxide injected into the pedal sinul of *Bullia* (Gastropoda: Prosobranchia). J Exp Biol 42: 509–519

Cheng TC (1981) Bivalves. In: Ratcliffe NA, Rowley AF (eds) Invertebrate blood cells, vol 1. Academic Press, London, pp 233–300

Cheng TC, Garrabrant TA (1977) Acid phosphatase in granulocytic capsules formed in strains of *Biomphalaria glabrata* totally and partially resistant to *Schistosoma mansoni*. Int J Parasitol 7: 467–472

Chien PK, Johnson PT, Holland ND, Chapman FA (1970) The coelomic elements of sea urchins (*Strongylocentrotus*). IV. Ultrastructure of the coelomocytes. Protoplasma 71: 419–442
Cohen N, Sigel MM (1982) The reticuloendothelial system. A comprehensive treatise, 3. Phylogeny and ontogeny. Plenum Press, New York
Cooper EL (1968) Transplantation immunity in annelids. I. Rejection of xenografts exchanged between *Lumbricus terrestris* and *Eisenia foetida*. Transplantation 6: 322–337
Cooper EL (1969) Specific tissue graft rejection in earthworms. Science 166: 1414–1415
Cooper EL (1976) Cellular recognition of allografts and xonografts in invertebrates. In: Marchalonis JJ (ed) Comparative immunology. Blackwell, Oxford, pp 36–79
Cooper EL (1992) Overview of immunoevolution. Boll Zool 59: 119–128
Cooper EL, Raftos DA, Kelly KL (1992a) Immunobiology of tunicates: the search for precursors of the vertebrate immune system. Boll Zool 59: 175–181
Cooper EL, Rinkevich B, Uhlenbruck G, Valempois P (1992b) Invertebrate immunity: another view point. Scand J Immunol. 35: 247–266
Cooper EL, Stein EA (1981) Oligochetes. In: Ratcliffe NA, Rowley AF (eds) Invertebrate blood cells, vol 1. Academic Press, London, pp. 75–140
Cooper EL, Leung MK, Suzuki MM, Vick K, Cated P, Stefano GB (1993) An enkephalin-like molecule in earthworm coelomic fluid modifies leukocyte behavior. Dev Comp Immunol 17: 201–209
Cowden RR, Curtis SK (1974) The octopus white body: an ultrastructural survey. In: Cooper EL (ed) Invertebrate immunology. Plenum Press, New York, pp 77–90
Cowden RR, Curtis SK (1981) Cephalopods. In: Ratcliffe NA, Rowley AF (eds) Invertebrate blood cells, vol 1. Academic Press, London, pp 301–323
Cuénot L (1891) Études sur le sang et les glandes lymphatiques dans la série animale (2°parties: Invertébrés). Arch Zool Exp Gén Ser 2 9: 13–90
Cuénot L (1905) L'organe phagocytaire des crustaces decapodes. Arch Zool Exp Gén Ser 4 3: 1–16
Dales RP (1961) The coelomic and peritoneal cell systems of some sabellid polychaetes. Q J Microsc Sci 102: 327–346
Dales RP, Dixon LRJ (1981) Polychetes. In: Ratcliffe NA, Rowley AF (eds) Invertebrate blood cells, vol 2. Academic Press, London, pp 35–74
David CN, Murphy S (1977) Characterization of interstitial stem cells in hydra by cloning. Dev Biol 58: 372–383
Deck JD, Hay ED, Revel JP (1966) Fine structure and origin of the tunic of *Perophora viridis*. J Morphol 120: 267–280
Decker JM, Elmholt A, Muchmore AV (1981) Spontaneous cytotoxicity mediated by invertebrate mononuclear cells toward normal and malignant vertebrate targets: inhibition by defined mono- and disaccharides. Cell Immunol 59: 161–170
De Leo G, Ptricolo E, Frittitta G (1981) Fine structure of the tunic of *Ciona intestinalis* L. II. Tunic morphology, cell distribution and the functional importance. Acta Zool (Stockh) 62: 259–271
Deno T (1987) Autonomous fluorescence of eggs of the ascidian *Ciona intestinalis*. J Exp Zool 241: 71–79
Dikkeboom R, van der Knaap WPW, Maaskant JJ, de Jonge AJR (1985) Different subpopulations of haemocytes in juvenile, adult and *Trichobilharzia ocellata*-infected *Lymnaea stagnalis*: a characterization using monoclonal antibodies. Z Parasitenkd 71: 815–819
Dikkeboom R, Tijnage MGH, van der Knaap WPW (1988) Monoclonal antibody recognized hemocyte subpopulations in juvenile and adult *Lymnaea stagnalis*: functional characteristics and lectin binding. Dev Comp Immunol 12: 17–32
Du Pasquier L (1992) Origin and evolution of the vertebrate immune system. APMIS 100: 383–392
Eckelbarger KJ (1976) Origin and development of the amoebocytes of *Nicolea zoostericola* (Polychaeta Terebellidae) with a discussion of their possible role in oogenesis. Mar Biol 36: 169–182

Edds KT (1977a) Microfilament bundles. I. Formation with uniform polarity. Exp Cell Res 108: 452–456
Edds KT (1977b) Dynamic aspects of filopodial formation by reorganization of microfilaments. J Cell Biol 73: 479–491
Edds KT (1980) The formation and elongation of filopodia during transformation of sea urchin coelomocytes. Cell Motility 1: 131–140
Endean R (1958) The coelomocytes of *Holothuria leucosphilota*. J Microsc Sci 99: 47–60
Endean R (1961) The test of ascidian, *Phallusia mammillata*. Q J Microsc Sci 102: 107–117
Ermak TH (1976) In: Wright RK, Cooper EL (eds) Phylogeny of thymus and bone marrow-bursa cells. Elsevier, Amsterdam, pp 45–56
Evans DL, Cooper EL (1990) Natural killer cells in ectothermic vertebrates: cytotoxic cells from different species have similar biological activities. Bioscience 40: 745–753
Ey PL, Jenkin CR (1982) Molecular basis of self/nonself discrimination in the invertebrata. In: Cohen N, Sigel MM (eds) The reticuloendothelial system. A comprehensive treatise, 3. Phylogeny and ontogeny. Plenum Press, New York, pp 321–391
Fontaine AR, Lambert P (1973) The fine structure of the haemocyte of the holothurian *Cucumaria miniata* (Brandt). Can J Zool 51: 323–332
Fontaine AR, Lambert P (1977) The fine structure of the leucocytes of the holothurian, *Cucumaria miniata*. Can J Zool 55: 1530–1544
Franceschi C, Cossarizza A, Monti D, Ottaviani E (1991) Cytotoxicity and immunocyte markers in cells from the freshwater snail *Planorbarius corneus* L. (Gastropoda: Pulmonata): implication for the evolution of natural killer cells. Eur J Immunol 21: 489–493
Fuke MT (1979) Studies on the coelomic cells of some japanese ascidians. Bull Mar Biol Stn Asamushi 16: 143–159
Fuke M (1980) "Contact reactions" between xenogeneic or allogeneic coelomic cells of solitary ascidians. Biol Bull 158: 304–315
Fuke M (1990) Self and nonself recognition in the solitary ascidian, *Halocynthia roretzi*. In: Marchalonis J, Reinish C (eds) Defence molecules. Alan R Liss, New York, pp 107–117
Fuke M, Fukumoto T (1993) Correlative fine structural behavioral, and histochemical analysis of ascidian blood cells. Acta Zool (Stock) 74: 61–71
Furuta E (1989) Primary culture of cells from the land slug. In: Mitsuhashi J (ed) Invertebrate cell system application, vol II. CRC Press, New York, pp 235–241
Furuta E, Shimozawa A (1983) Primary culture of cells from the foot and mantle of the slug, *Incilaria fruhstorferi* Collinge. Zool Mag 92: 290–296
Furuta E, Shimozawa A (1994) The blood cell-producing site in lang slug, *Incilaria fruhstorferi*. Acta Anat Nippon 69: 751–764
Furuta E, Yamaguchi K, Aikawa M, Shimozawa A (1987) Phagocytosis by hemolymph cells of the land slug, *Incilaria fruhstorferi* Collinge (Gastropoda: Pulmonata). Anat Anz Jena 163: 89–99
George WC (1930) The histology of the blood of some Bermuda ascidians. J Morph Physiol 49: 385–413
George WC (1939) A comparative study of the blood of the tunicates. Q J Microsc Sci 81: 391–431
Hetzel HR (1963) Studies on Holothurian coelomocytes. I. A survey of coelomocyte types. Biol Bull 125: 289–301
Hetzel HR (1965) Studies on holothurian coelomocytes. II. The origin of coelomocytes and formation of brown bodies. Biol Bull 128: 102–111
Hildemann WH, Bigger CH, Johnston IS, Jokiel PL (1980) Characteristics of transplantation immunity in the sponge, *Callyspongia diffusa*. Transplantation 30: 362–367
Hirose E, Saito Y, Watanabe H (1988) A new type of the manifestation of colony specificity in the compound ascidian, *Botrylloides violaceus* Oka. Biol Bull 175: 240–245
Hirose E, Saito Y, Watanabe H (1991) Tunic cell morphology and classification in botryllid ascidians. Zool Sci 8: 951–958
Hose JE, Martin GG, Nguyen VA, Jucas J, Rosenstein T (1987) Cytochemical features of shrimp hemocytes. Biol Bull 173: 178–187

Hose JE, Martin GG, Gerard AS (1990) A decapod hemocyte classification scheme integrating morphology, cytochemistry, and function. Biol Bull 178: 33–45
Hughes TK Jr, Smith EM, Leung MK, Stefano GB (1992) Evidence for the conservation of an immunoreactive monokine network in invertebrates. Ann NY Acad Sci 650: 74–80
Ito T, Matsutani T, Mori K, Nomura T (1992) Phagocytosis and hydrogen peroxide production by phagocytes of the sea urchin *Strongylocentrotus nudus*. Dev Comp Immunol 16: 287–294
Janeway CA (1989) Natural killer cells. A primitive immune system. Nature 341: 108
Jackson AD, Smith VJ, Peddie CM (1993) In vitro phenoloxidase activity in the blood of *Ciona intestinalis* and other ascidians. Dev Comp Immunol 177: 97–108
Johnson PT (1969a) The coelomic elements of sea urchins (*Strongylocentrotus*). J Invertebr Pathol 13: 25–41
Johnson PT (1969b) The coelomic elements of sea urchins (*Strongylocentrotus*). II. Cytochemistry of the coelomocytes. Histochemie 17: 213–231
Johnson PT (1969c) The coelomic elements of sea urchins (*Stronglylocentrotus*). III. In vitro reaction to bacteria. J Invertebr Pathol 13: 42–62
Johnson PT, Beeson RJ (1966) In vitro studies on *Patria miniata* (Brandt) coelomocytes, with remarks on revolving cysts. Life Sci 5: 1641–1666
Johnson PT (1987) A review of fixed phagocytic and pinocytotic cells of decapod crustaceans, with remarks of hemocytes. Dev Comp Immunol 11: 679–704
Jones JC (1962) Current concepts concerning insect hemocytes. Am Zool 2: 209–246
Kalk M (1963) Intracellular sites of activity in the histogenesis of tunicate vanadocytes. Q J Microsc Sci 104: 483–493
Kelly KL, Cooper EL, Raftos DA (1992) In vitro allogeneic cytotoxicity in the solitary urochordates. J Exp Zool 262: 202–208
Kelly KL, Cooper EL, Raftos DA (1993) A humoral opsonin from the solitary urochordate *Styela clava*. Dev Comp Immunol 17: 29–39
Klein J (1989) Are invertebrates capable of anticipatory immune responses? Scand J Immunol 29: 499–505
Kondo M, Itami T, Takahashe Y, Fujii R, Tomonaga S (1994) Structure and function of the lymphoid organ in the kuruma prawn. Dev Comp Immunol 18 (Suppl): S 109
Lai-Fook J (1973) The structure of the haemocytes of *Calpodes ethlius* (Lepidoptera). J Morphol 139: 79–104
Landureau JC (1968) Cultures in vitro de cellules embryonnaires de blattes (insectes dictyopteres). II. Obtention de lignees cellulaires a multiplication continue. Exp Cell Res 50: 323–337
Landureau JC, Grellet P (1975) Obtention de lingnees permanentes d'hemocytes de blatte: caracteristiques physiologiques et ultrastructurales. J Insect Physiol 21: 137–151
Leclerc M, Brillouet C, Luquet G, Binaghi RA (1986) Production of an antibody-like factor in the sea star *Asterias rubens*: involvement of at least three cellular populations. Immunology 57: 479–482
Lehn HZ (1951) Teilungsfolgen und Determination von I-zellen fuer die Cnidenbildung bei Hydra. Z Naturforsch B 6: 388–391
Martin GG, Hose JE, Kim JJ (1987) Structure of hematopoietic nodules in the ridgeback prawn, *Sicyonia ingentis*: light and electron microscopic observations. J Morphol 192: 193–204
Michibata H, Hirata J, Terada T, Sakurai H (1988) Autonomous fluorescence of ascidian blood cells with special reference to identification of vanadocytes. Experientia 44: 906–907
Michibata H, Uyama T, Hirata J (1990) Vanadium-containing blood cells (Vanadocytes) show no fluorescence due to the tunichrome in the ascidian *Ascidia sydneiensis samea*. Zool Sci 7: 55–61
Milanesi C, Burighel P (1978) Blood cell ultrastructure of the ascidian *Botryllus schlosseri*. I. Hemoblast, granulocytes, macrophage, morula cell and nephrocyte. Acta Zool (Stockh) 59: 135–147
Morgan TH (1938) The genetic and the physiological problems of self-sterility in Ciona. 1. Data on self- and cross-fertilization. J Exp Zool 78: 271–318

Mürer EH, Levin J, Holme R (1975) Isolation and studies of the granules of the amoebocytes of *Limulus polyphemus*, the horseshoe crab. J Cell Physiol 86: 533–542
Muta T, Hashimoto R, Miyata T, Nishimura H, Toh Y, Iwanaga S (1990) Proclotting enzyme from horseshoe crab hemocytes. cDNA cloning, disulfide locations and subcellular localization. J Biol Chem 265: 22426–22433
Nakamura T, Morita T, Iwanaga S (1985) Intracellular proclotting enzyme in Limulus (*Tachypleus tridentatus*) hemocytes: its purification and properties. J Biochem 97: 1561–1574
Ohtake S, Shishikura F, Tanaka K (1989) Roles of granular amoebocytes on adhesion/aggregation of hemolymph of *Halocynthia roretzi*. Zool Sci (Abstr) 6: 1106
Ohuye T (1936) On the coelomic corpuscles in the body fluid of some invertebrates. III. The histology of the blood of some Japanese ascidians. Sci Rep Tohoku Univ 11: 191–206
Oka M (1969) Studies on *Penaeus orientalis* Kishinoueye-VIII Structure of the newly found lymphoid organ. Bull Jpn Soc Sci Fish 35: 245–250
Otto JJ, Kane RE, Bryan J (1979) Formation of filopodia in coelomocytes: localization of fascin, a 58,000 dalton actin cross-linking protein. Cell 17: 285–293
Overton J (1966) The fine structure of blood cells in the ascidian *Perophora viridis*. J Morphol 119: 305–326
Parinello N, Patricolo E, Canicatti C (1977) Tunicate immunobiology. I. Tunic reaction of *Ciona intestinalis* L. to erythrocyte injection. Boll Zool 44: 373–381
Parrinello N, Patricolo E, Canicatti C (1984) Inflammatory-like reaction in the tunic of *Ciona intestinalis* (Tunicata). I. Encapsulation and tissue injury. Biol Bull 167: 229–237
Parinello N, Arizza V, Cammarata M, Parinello DM (1993) Cytotoxic activity of *Ciona intestinalis* (Tunicata) hemocytes: properties of the in vitro reaction against erythrocyte targets. Dev Comp Immunol 17: 19–27
Porchet-Hennere E, Dugimont T, Fischer A (1992) Natural killer cells in a lower invertebrates *Nereis diversicolor*. Eur J Cell Biol 58: 99–107
Price CD, Ratcliffe NA (1974) A reappraisal of insect haemocyte classification by the examination of blood from fifteen insect orders. Z Zellforsch Mikrosk Anat 147: 537–549
Raftos DA, Cooper EL (1991) Proliferation of lymphocyte-like cells from the solitary tunicate, *Styela clava*, in response to allogeneic stimuli. J Exp Zool 260: 391–400
Raftos DA, Tait NN, Briscoe DA (1987) Cellular basis of allograft rejection in the solitary urochordata, *Styela plicata*. Dev Comp Immunol 11: 713–725
Raftos DA, Stillman DL, Cooper EL (1990) In vitro culture of tissue from the tunicate *Styela clava*. In Vitro Cell Dev Biol 26: 962–970
Raftos DA, Cooper EL, Habicht GS, Beck G (1991) Invertebrate cytokines – tunicate cell proliferation stimulated by an interleukin-1-like molecule. Proc Natl Acad Sci USA 88: 9518–9522
Ratcliffe NA, Rowley AF (1979) A comparative synopsis of the structure and function of the blood cells of insects and other invertebrates. Dev Comp Immunol 3: 189–221
Ratcliffe NA, Rowley AF (1981a) Invertebrate blood cells 1. Academic Press, London
Ratcliffe NA, Rowley AF (1981b) Invertebrate blood cells 2. Academic Press, London
Ratcliffe NA, Rowley AF, Fitzgerald RW, Rhodes CP (1985) Invertebrate immunity: basic concepts and recent advances. Int Rev Cytol 97: 183–354
Rizki TM (1962) Experimental analysis of hemocyte morphology in insects. Am Zool 2: 247–256
Roch P, Valembois P (1978) Evidence for concanavalin A-receptors and their redistribution on lumbricid leukocytes. Dev Comp Immunol 2: 51–63
Rowley AF (1977) The role of the haemocytes of *Clitumnus extradentatus*, in haemolymph coagulation. Cell Tissues. Res 182: 513–524
Rowley AF, Ratcliffe NA (1976) The granular cells of *Galleria mellonella* during clotting and phagocytic reactions in vitro. Tissue Cell 8: 437–446
Rowley AF, Ratcliffe NA (1981) Insect. In: Ratcliffe NA, Rowley AF (eds) Invertebrate blood cells, vol 2. Academic Press, London, pp 421–488
Sato S, Akai H, Sawada H (1976) An ultrastructural study of capsule formation by *Bombyx* hemocytes. Annot Zool Jpn 49: 177–188

Sawada T, Fujikura Y, Tomonaga S, Fukumoto T (1991) Classification and characterization of ten hemocyte types in the tunicate *Halocynthia roretzi*. Zool Sci 8: 939–950
Sawada T, Zhang J, Cooper EL (1993) Classification and characterization of hemocytes in *Styela clava*. Biol Bull 184: 87–96
Swada T, Zhang J, Cooper EL (1994) Sustained viability and proliferation of hemocytes from the cultured pharynx of *Styela clava*. Mar Biol 119: 597–603
Schmit AR, Ratcliffe NA (1978) The encapsulation of araldite implants and recognition of foreignness in *Clitumnus extradentatus*. J Insect Physiol 24: 511–521
Sminia T (1981) Gastropods. In: Ratcliffe NA, Rowley AF, (eds) Invertebrate blood cells, vol 2. Academic Press, London, pp 191–232
Sminia T, Borghart-Reinders E, van de Linde AW (1974) Encapsulation of foreign materials experimentally introduced into the fresh water snail *Lymnaea stagnalis*. Cell Tissue Res 153: 307–326
Smith VJ (1981) The echinoderms. In: Ratcliffe NA, Rowley AF (eds) Invertebrate blood cells, vol 2. Academic Press, London, pp 513–562
Sohi SS (1979) Hemocyte cultures and insect hemocytology. In: Gupta AP (ed) Insect hemocytes. Cambridge Univ Press, Cambridge, pp 259–318
Stefano GB, Leung MK, Zhao X, Scharrer B (1989) Evidence for the involvement of opioid neuropeptides in the adherence and migration of immunocompetent invertebrate hemocytes. Proc Natl Acad Sci USA 86: 626–630
Stefano GB, Shipp MA, Scharrer B (1991) A possible immunoregulatory function for [Met]-enkephalin-Arg^6-Phe^7 involving human and invertebrate granulocytes. J Neuroimmunol 31: 97–103
Stein EA, Cooper EL (1978) Cytochemical observations of coelomocytes from the earth-worm *Lumbricus terrestris*. Histochem J 10: 657–678
Stein EA, Avtalion RR, Cooper EL (1977) The coelomocytes of the earthworm *Lumbricus terrestris*: morphology and phagocytic properties. J Morphol 153: 467–477
Taneda Y, Saito Y, Watanabe H (1985) Self and non-self discrimination in ascidians. Zool Sci 2: 433–442
Terwilliger NB, Terwilliger RC, Schabtach E (1985) In: Cohen WD (ed) Blood cells of marine invertebrates. Alan R Liss, New York, pp 193–225
Teschemacher H, Koch G, Scheffler H, Hildebrand A, Brantl V (1990) Opioid peptides: immunological significance? Ann NY Acad Sci 594: 66–77
Toh Y, Mizutani A, Tokunaga F, Muta T, Iwanaga S (1991) Morphology of the granular hemocytes of the Japanese horseshoe crab *Tachypleus tridentatus* and immunocytochemical localization of clotting factors and antimicrobial substances. Cell Tissue Res 266: 137–147
Tyson CJ, Jenkin CR (1974) The cytotoxic effect of haemocytes from the crayfish *Parachaeraps bicarinatus* on tumor cells of vertebrates. Aust J Exp Biol Xed Sci 52: 915–923
Valembois P, Boiledieu D (1980) Fine structure and functions of haemerythrocytes and leukocytes of *Sipunculus nudus*. J Morph 163: 69–77
Valembois P, Lassegues M, Roch P (1992) Formation of brown bodies in the coelomic cavity of the earthworm *Eisenia andrei* and attendant changes in shape and adhesive capacity of constitutive cells. Dev Comp Immunol 16: 95–101
Vallee JA (1967) Studies of the blood of *Ascidia nigra* (Savigny). I. Total blood cell counts, differential blood cell counts, and hematocrit values. Bull South Calif Acad Sci 66: 23–28
Van de Vyver G (1981) Organisms without special circulatory systems. In: Ratcliffe NA, Rowley AF (eds) Invertebrate blood cells, vol 1. Academic Press, London, pp 19–32
Van Praet M, Doumenc D (1974) Morphologie et morphogenese experimentale dù tentacule chez *Actina equina* L. J Microsc Biol Cell 23: 29–38
Vethamany VG, Fung M (1971) The fine structure of coelomocytes of the sea urchin *Strongylocentrotus drobachiensis* (Muller O. F.). Can J Zool 50: 77–81
Wago H (1982) Involvement of microfilaments in filopodial function of phagocytic granular cells of the silkworm, *Bombyx mori*. Dev Comp Immunol 6: 655–664
Watson GM, Mariscal RN (1983a) The development of a sea anemone tentacle specialized for aggression: morphogenesis and regression of the catch tentacle of *Haliplanella luciae* (Cnidaria, Anthozoa). Biol Bull 164: 506–517

Watson GM, Mariscal RN (1983b) Comparative ultrastructure of catch tentacles and feeding tentacles in the sea anemone *Haliplanella*. Tissue Cell 15: 939–953

Wittke M, Renwrants L (1984) Quantification of cytotoxic hemocytes of *Mytilus edulis* using a cytotoxicity assay in agar. J Invertebr Pathol 43: 248–253

Wood PJ, Visentin LP (1967) Histological and histochemical observations on hemolymph cells in the crayfish, *Orconectes virilis*. J Morphol 123: 559–568

Wright RK (1981) Urochordates. In: Ratcliffe NA, Rowley AF (eds) Invertebrate blood cells, vol 2. Academic Press, London, pp 565–626

Wright RK, Cooper EL (1975) Immunological maturation in the tunicate *Ciona intestinalis*. Am Zool 15: 21–27

Yoshida R, Takikawa O, Oku T, Habara-Ohkubo A (1991) Mononuclear phagocytes: a major population of effector cells responsible for rejection of allografted tumor cells in mice. Proc Natl Acad Sci USA 88: 1526–1530

Yoshino TP (1976) The ultrastructure of the circulating hemolymph cells of the marine snail *Cerithidea californica* (Gastropoda; Prosobranchiata) J Morphol 150: 485–494

Yoshino TP, Granath Jr WO (1983) Identification of antigenically distinct hemocyte subpopulations in *Biomphalaria glabrata* (Gastropoda) using monoclonal antibodies to surface membrane markers. Cell Tissue Res 232: 553–564

Yoshino TP, Granath Jr WO (1985) Surface antigens of *Biomphalaria glabrata* (Gastropoda) hemocytes: functional heterogeneity in cell subpopulations recognized by a monoclonal antibody. J Invertebr Pathol 45: 174–186

Zaniolo G, Trentin P (1987) Regeneration of the tunic in the colonial ascidian, *Botryllus schlosseri*. Acta Embryol Morphol Exp 8: 173–180

Zhang H, Sawada T, Cooper EL, Tomonaga S (1992) Electron microscopic analysis of tunicate (*Halocynthia roretzi*) hemocytes. Zool Sci 9: 551–562

Chapter 2

Mechanisms of Antigen Processing in Invertebrates: Are There Receptors?*

L. Tučková and M. Bilej

Contents

* Dedicated to Jaroslav Rejnek

Department of Immunology, Institute of Microbiology, Academy of Sciences of the Czech Republic, Vídeňská 1083, 142 20 Prague 4, Czech Republic

Advances in Comparative and Environmental Physiology, Vol. 23

1 Introduction

1.1 Immunodefense

The ability to recognize *self* and *non-self* exists in all animal species. Unicellular animals, such as protozoans, which often engulf living microorganisms, must discriminate between them and nutrition proteins, to prevent damage to their own proteins during digestive processes. The mechanism of discrimination at this level is unknown. One can assume that the specificity is based on substrate specificity of the proteolytic enzymes (Valembois et al. 1973; Ratcliffe et al. 1984, 1991; Tučková et al. 1986a). The main defense mechanisms in multicellular invertebrates are certainly represented by innate factors. Microorganisms that break the outer protective barrier and invade the host are mainly eliminated by phagocytosis which can be potentiated by humoral factors. Moreover, body fluids (e.g. hemolymph, celomic fluid) contain antibacterial molecules that probably prevent the multiplication of these bacteria.

Together with the evolution of multicellular organisms the necessity to recognize *self* and *non-self* arose to prevent the undesirable intrusion of metabolically different cells that originate from other potentially pathogenic multicellular organisms; they could cause serious damage to the host. Within the metazoa, sea sponges in particular developed marked histocompatibility polymorphism that equips them to reject xenografts by means of cytotoxic reactions accompanied by short-term immune memory (Cooper 1970; Hildemann 1981). This suggests that their effector cells and/or products can reproduce after an initial response.

Molecules such as agglutinins, lysins, and other antibacterial factors take part in natural resistance. For example, agglutinins are mostly lectins that are capable of interacting with specific soluble or membrane bound sugar molecules. Since lectins are widely distributed and act like agglutinins, opsonins, and mitogens, one can assume that discrimination in which sugar moieties are involved was crucial during evolution and remains functionally important (Stein et al. 1982; Wojdani et al. 1982; Komano et al. 1987, etc.)

According to our present knowledge, it is also of interest to study the induction of molecules with the ability to interact with stimulating antigens other than those associated with viable microorganisms. The adaptive response was well documented especially in two invertebrate taxons: insects and annelids.

1.2 Humoral Responses in Selected Invertebrates

Humoral responses in insects can be induced by injecting them with either live, non-pathogenic bacteria or heat-killed pathogens. The response is characterized by the product of several factors not present in naive (non-stimulated) hemolymph. Induction has been studied intensively in *Galleria mellonella*, *Hyalophora cecropia*, and *Sarcophaga peregrina* and to a lesser extent in other species. Whereas in *Galleria*, lysozyme, which is bactericidal for only a few Gram-positive microorganisms, has been purified, in the other species a set of newly synthesized, antibacterial proteins has been discovered (Boman and Steiner 1981; Götz and Trenczek 1991).

In *Hyalophora cecropia* this set includes several antibacterial proteins: lysozyme, *cecropins*, *attacins*, and *hemolin*. These proteins can also be induced to a lesser extent by wounding or by injecting them with components of the bacterial cell wall, such as lipopolysaccharides or the degradative products from peptidoglycans. Antibacterial activity, whose specificity seems to be broad begins to be expressed in the hemolymph after a lag period of 8–10 h, rises up to 8 days and then gradually declines. The three principal *cecropins* A, B and D are small basic proteins (4 kDa) with a comparatively hydrophobic region. The *attacins* are larger (20 kDa) and are expressed in two main forms: basic, and acidic or neutral. *Hemolin*, still larger, is a 48-kDa protein present in low but significant amounts in the hemolymph of *H. cecropia* pupae. After injecting these pupae with live bacteria the concentration of hemolin increases up to 18 times. According to functional analyses, *hemolin* is one of the first hemolymph components that binds to bacterial surfaces and it probably takes part in the active formation of a protein complex that is likely to initiate antibacterial responses. A cDNA bank has been prepared and complete amino acid sequence for lysozyme, *cecropins*, *attacin* and *hemolin* determined. It is of interest that *hemolin* has been included as a member of the Ig superfamily (Boman et al. 1985; Sun et al. 1990).

In *Sarcophaga peregrina* a group of humoral, antibacterial proteins, termed *sarcotoxin* I, II, and III, have been isolated and sequenced. *Sarcotoxin* I consists of three proteins (IA, IB, and IC) with an almost identical primary structure that comprises 39 amino acid residues and differs only in 2–3 amino acids. *Sarcotoxin* II is also a mixture of three structurally related proteins (IIA, IIB, and IIC; 24 kDa). The cDNA analyses have revealed that *sarcotoxin* I and *cecropins* belong to a similar protein family with antibacterial activity. *Sarcotoxin* III is a single glycine-rich protein (7 kDa) induced in the hemolymph in response to injury of the larval body wall.

In fact, injury to the body wall of larval or adult *Sarcophaga peregrina* stimulates the induction not only of antibacterial activity but also increased levels of agglutinating proteins, i.e., the production of "sarcophaga lectin". These proteins participate in insect defense mechanisms and are expected to increase under conditions of injury in order to eliminate pathogens that invade the host through the damaged body wall (Baba et al. 1987; Komano et al. 1987; Ando and Natori 1988).

In hemolymph of stimulated larvae of the dipteran, *Phormia terranovae*, a family of peptides—*diptericins* (8 kDa) appears. They are active against Gram-negative bacteria only, together with cecropin-like and perhaps other, not yet identified, antibacterial peptides. The immune hemolymph of *Phormia* does not contain lysozyme and its potent activity against Gram-positive bacteria is attributed to two positively charged peptides (4 kDa). These peptides, structurally related to microbicidal, cationic peptides derived from mammalian granulocytes (defensins) are designated "*insect defensins*" (Lambert et al. 1989).

Another family of inducible peptides has been isolated from both larval and adult honeybees (*Apis mellifera*) and referred to as *apidaecins*. These heat-stable, non-helical peptides (2100 daltons) are active against a wide range of plant-associated bacteria and some human pathogens by means of a bacteriostatic, rather than a lytic process. Chemically synthesized *apidaecins* display the same bactericidal activity as natural counterparts. The active antibacterial peptides are detectable only in the lymph of adult honeybees, whereas larvae contain their inactive precursor molecules (Casteels et al. 1989). Adult American cockroaches (*Periplaneta americana*) generate inducible protective responses to bacteria and soluble, protein toxins. The response to bacteria is apparently a biphasic one in that the first phase is acute and non-specific, whereas the second is longer and specific (Karp 1985; Faulhaber and Karp 1991).

While interest in the biology of insects is absolutely clear owing to their importance in agriculture, ecology, epidemiology, industry, etc., the importance of annelids has not yet received as much attention. It should be mentioned that terrestrial oligochaetes represent an excellent, ecological monitor from several viewpoints. First, there is the maintenance of soil fertility. Second, trials on the application of biologically active compounds isolated from earthworms have recently been performed (Mihara et al. 1991; Nagasawa et al. 1991; Hrženjak et al. 1992). Third, earthworms have been used as experimental sentinels for detecting, by changes in the immune system, the effects of environmental pollution (see Goven and Kennedy, Chap. 6, Vol. 24). The interest of comparative immunologists has been focused on oligochaetes since the early1960s when the first results indicating the ability of earthworms to reject foreign grafts were published (Valembois 1963; Cooper 1965). Ten years ago we aimed our experimental efforts at following some aspects of *self/non-self* recognition in the earthworms *Lumbricus terrestris* and *Eisenia foetida* after parenteral stimulation using a complex of hapten coupled to a protein carrier.

We realize that our review is somewhat incomplete since it focuses on only one invertebrate taxon (earthworms: *Oligochaeta*, *Annelida*) traditionally studied by our group, nevertheless, we have tried to outline some characteristics of adaptive responses with special emphasis on antigen processing and the origin of receptor molecules involved in invertebrate defense.

2 General Armamentarium of Earthworm Natural Resistance

2.1 Morpho–Functional Features

Earthworms are endowed with a true celomic cavity of mesenchymal origin that is filled by the celomic fluid containing free, wandering celomocytes derived from the celomic lining. The celomic cavity communicates with the outer environment by dorsal pores through which metabolites are excreted by paired nephridia. The celomic cavity is segmented with transversal septa and regulated transport of celomic fluid and celomocytes between neighboring segments is provided by channels comprised sphincters within the septa.

2.1.1 Dorsal Pores

The first barrier that protects annelids from infection is certainly the abundance of secreted mucus containing agglutinins (Porchet-Henneré and M'Berri 1987). Nevertheless, it should be noted that the celomic cavity is not aseptic and that it always contains bacteria invading from the outer environment mainly via dorsal pores. On the other hand, the dorsal pores represent one of the important routes for the elimination of bacteria. Dales and Kalaç (1992) reported that the celomic fluid contains approximately 6×10^5/ml of naturally occurring bacteria, while the number of potentially phagocytic cells is more than ten times higher. Under normal conditions, the excess of phagocytes combined with the humoral antibacterial factors could easily prevent the celomic bacteria from multiplying. The phagocytes that become "exhausted" for further uptake are eliminated by expulsion through dorsal pores. Thus one can envision continuous exchange of foreign material between the celom and the outer environment. This was documented by revealing the presence of foreign corpuscles that had been introduced artificially into the celomic cavity on the surface of the body and vice versa, i.e., corpuscles present in the outer environment have been found in the celomic fluid (Cameron 1932). Furthermore, under conditions of stress the celomic fluid can be rapidly expelled by increased intra-celomic pressure (Fig. 1).

2.1.2 Nephridia

Another natural way of eliminating foreign particles from the body is effected by nephridia (Cameron 1932). Soluble dyes injected into the celomic cavity are almost immediately excreted by nephridia as revealed by staining their components: the nephrostome and middle-tube cells were found to be both excretory and phagocytic (Villaro et al. 1985).

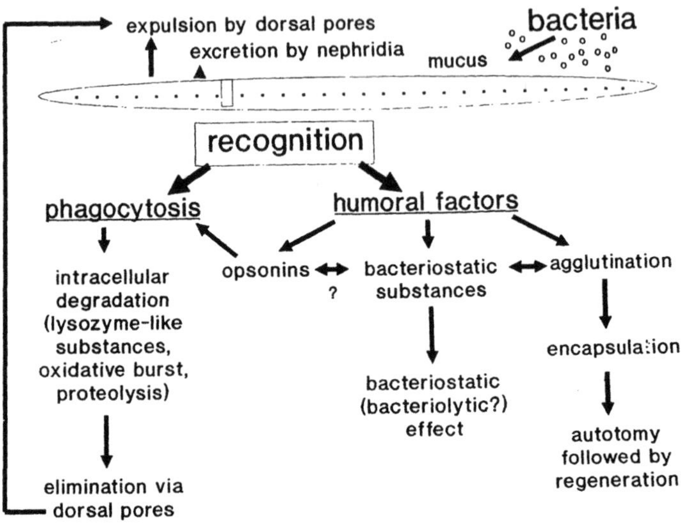

Fig. 1. General armamentarium of natural resistance of the earthworm. The first protective barrier is represented by secreted mucus containing agglutinins. Bacteria invading the celomic cavity are directly expelled via dorsal pores and excreted by nephridia or can be engulfed by celomocytes. Moreover, humoral factors are involved in their elimination: Agglutinated bacteria can be encapsulated, antibacterial substances prevent bacteria from multiplying, opsonins facilitate phagocytosis. All humoral factors involved in antibacterial defense seem to be connected in complex protective system

2.2 Encapsulation

Large foreign bodies like parasites or agglutinated bacteria are eliminated by encapsulation. This process begins like phagocytosis with recognition, but phagocytosis does not occur because of the size of the objects (Bang 1973). Foreign bodies are surrounded by all types of celomocytes within 24 h and after several days a dense capsule, composed of flattened cells, is formed. The process begins first by small vessels which grow outwards from the body wall and are surrounded by proliferating cells. Later, thin strands of fibrous tissue appear around the foreign object forming a fibrous capsule (sometimes called "brown body"), owing to the content of brown pigment, melanin (Ratcliffe et al. 1985) which is often calcified after the degenerative process is completed (Cameron 1932).

The process of encapsulation in *Eisenia foetida andrei* has been recently described in detail by Valembois et al. (1992). They propose using the term "brown body" only for terminal stages characterized by deposits of melanin, while for smaller "immature" structures, they suggest using the term "nodule" or "aggregate". When a convex disk-shaped mass of celomocytes surrounding an object recognized as *non-self* reaches the diameter 1–2 mm, its external cells flatten and lose their adhesiveness toward free celomocytes. The nodule then becomes distinctly brown and migrates to the posterior segments of the celomic cavity where it can be eliminated by autotomy (Keilin 1925). The process of autotomy of caudal segments, similar to wound healing, also known to be well developed in annelids, seems to be under neurohormonal control (Herlant-Meewis and Deligne 1964; Herlant-Meewis 1966; Golding 1974; Olive 1974; Alonso-Bedate and Sequeros 1985). Following the healing process, a rapid and efficient regeneration of the damaged area occurs. Examination of nodules present in *E. f. andrei* (Valembois et al. 1992) revealed that most nodules (85%) contained tissue wastes, especially necrotic muscle cells and setae. In 29% of nodules agglutinated bacteria were observed (isolated bacteria are phagocytized) and in 9% of nodules gregarines and nematodes were present. In some cases, two or three types of object were observed in the same aggregate. Similar results were described in some polychaetes especially in work which attempts to explain host–parasite interactions (Vivier and Henneré 1964; Schrevel 1969, 1970, 1971; Porchet-Henneré and M'Berri 1987), nevertheless, the exact mechanism remains to be solved.

2.3 Phagocytosis

2.3.1 General Characteristics

Phagocytosis undoubtedly plays a key role in natural cellular defense in earthworms. Though the phagocytic properties of celomocytes were already mentioned by Metchnikoff in 1887, the first detailed description was published by Cameron (1932). He observed phagocytosis of numerous inert particles, cells, and bacteria, and found that all types of celomocytes, with the exception of chloragogen cells, were phagocytic, but that their activity differed. While inert particles were engulfed mainly by granulocytes, bacteria and foreign cells were phagocytized by basophilic and neutrophilic cells. In contrast the phagocytic index (i.e., number of particles per single cell) was highest in basophils and neutrophils even when the engulfment of inert particles was analyzed. Phagocytosis commenced almost at once and proceeded rapidly. Only the ingestion of several particular bacterial strains was delayed, a result that could be due to the production of soluble exotoxins that might exert an inhibitory effect on phagocytosis. Cameron's results were confirmed later in reports describing uptake of inert particles (Stein et al. 1977; Bilej et al. 1990a, 1991a), bacteria (Dales and Kalaç 1992), yeast (Stein and Cooper 1981), and erythrocytes (Laulan et al. 1988).

Phagocytosis begins by the recognition of *non-self* which is followed by the engulfment and destruction of the particles. Particularly interesting evidence which supports this presumption was revealed by experiments that analyzed the uptake of spermatozoa (Cameron 1932). Isolated spermatozoa derived from mice, rats, rabbits, and mature worms (*Lumbricus terrestris*, *Allolobophora longa*, and *Octolasium cyaneum*) were injected into the celomic cavity of *L. terrestris*. While xenogeneic spermatozoa were phagocytized, mainly by basophilic celomocytes within 24 h starting with adherence and agglutination, homologous spermatozoa survived in the celomic cavity for 3 or 4 days without any uptake; similar results were obtained in vitro.

2.3.2 Opsonization

The engulfment of foreign materials by phagocytic cells may be generally enhanced by humoral factors – opsonins – that coat the phagocytized particle. The opsonins are then recognized by receptors on the surfaces of phagocytic cells thus facilitating the uptake. Although no effect of celomic fluid factors on phagocytosis of sheep erythrocytes was observed (Cooper 1973), the presence of celomic fluid influenced the uptake of yeast, *Saccharomyces cerevisiae* (Stein and Cooper 1981). Surprisingly, whereas the phagocytic activity of neutrophilic cells was depressed in whole celomic fluid, it was significantly increased in diluted fluid. Moreover, the phagocytic activity of neutrophils was augmented when yeast had been preincubated i.e., opsonized, in celomic fluid. A possible explanation for this is that the effect of opsonization could be masked because of the relatively low frequency of neutrophils in the whole celomocyte population (about 20%) compared to that of other unaffected cell types.

The opsonizing effect of components of celomic fluid has also been reported for the uptake of synthetic 2-hydroxyethylmethacrylate particles (HEMA; Bilej et al. 1990a, 1991a). Preincubation of HEMA particles in celomic fluid resulted in a significant increase in both in vivo and in vitro phagocytic activity and the resulting phagocytic index. By flow cytometry, using fluorescein-labeled HEMA particles, two subpopulations differing in phagocytic indexes were identified. One subpopulation with the higher phagocytic index only appeared when opsonized particles were tested or when the cells collected from antigen-stimulated worms were used (Bilej et al. 1991a). As necessary proof of their properties, the proteins in celomic fluid of *L. terrestris* adsorbed on HEMA particles were eluted and subjected to SDS-PAGE (Bilej et al. 1990a). Analysis of eluted proteins revealed the presence of two protein bands corresponding to molecular weights of 62 and 68 kDa.

In *E. foetida* the opsonizing function of celomic fluid proteins seems to be connected with their role in hemolytic processes (Šinkora et al. 1993). The celomic fluid used for preincubation of HEMA particles has lower hemolytic activity when compared to controls. Although proteins adsorbed on HEMA surfaces did not exert any hemolysis, they did compensate for the decreased hemolytic capacity of "opsonizing" celomic fluid almost to the level of control values or they increased the hemolytic capacity of control celomic fluid. SDS-PAGE analyses showed that by eluting with borate buffer, a broad spectrum of adsorbed proteins with molecular weights of 25 to 88 kDa was detected. After borate/glycine HCl III elution, only three protein bands were obtained—38, 40, and 51 kDa. Due to the fact that both eluates had compensating effects we suggest that at least one of the three adsorbed proteins is involved in the hemolytic cascade (see Sect. 2.4.1).

2.3.3 Degradation of Engulfed Material

Several mechanisms of eliminating phagocytized material have already been mentioned earlier: celomocytes that become "exhausted" are expelled via dorsal pores (Sect. 2.1.1), excretion and phagocytosis were observed in the middle tube of nephridia (Sect. 2.1.2), larger bodies or agglutinated small bodies (like bacteria) can be neutralized by encapsulation (Sect. 2.2). Furthermore, there are at least three pathways of intracellular degradation: the effect of proteolytic enzymes and lysozyme-like substances will be detailed later (Sects. 2.5 and 2.4.2). The third mechanism represents the induction of an oxidative burst, i.e., the production of highly reactive oxygen radicals that represent an effective mechanism of intracellular killing of bacteria. Though the oxidative activity has been detected both in the celomic fluid and in chloragocyte vesicles called chloragosomes (Valembois et al. 1991), the oxidative burst was not triggered during the phagocytosis of inert, synthetic particles (Bilej et al. 1991b). Based on these results it can be hypothesized that celomocytes discriminate between antigenic and non-antigenic materials, and killing reactive oxygen radicals are produced only when the uptake of antigenic (usually living) particles commences.

2.4 Humoral Factors of Natural Defense

Although the exact mechanism of antibacterial and hemolytic activities are detailed by Roch (Chap. 4, this Vol.), it will be useful to mention some general aspects of humoral defense in the following paragraphs.

2.4.1 Antibacterial Molecules with Hemolytic Properties

When experiments studying the effects of celomic fluid on common vertebrate or human bacteria revealed no positive results (Cooper et al. 1969), Valembois et al. (1982) focused their interest on bacteria habitually associated with the earthworm's biotope. They isolated 23 strains of telluric bacteria from manure and exposed them to the celomic fluid of *Eisenia foetida andrei*. Only six bacterial strains, both Gram-positive and Gram-negative, manifesting high pathogenicity when inoculated into the celomic cavity, were sensitive to the action of the celomic fluid. All sensitive bacteria expressed at least one superficial antigen common to vertebrate erythrocytes but distinct from the Forssman antigen (Roch et al. 1981). Owing to these results and to the thermolability of antibacterial activity, it was suggested that at least one of the bacteriostatic substances belongs to the hemolytic system (Valembois et al. 1982). When the celomic fluid was subjected to gel filtration, 11 fractions were obtained, 7 of which were characterized by significant antibacterial activity. At least four of the isolated fractions were connected with hemolytic or hemagglutinating activities (175-kDa pentamer of 35-kDa hemagglutinating subunits, Roch et al. 1984; 20-, 40-, and 45-kDa molecules possessed either hemolytic or hemagglutinating activities,

Vaillier et al. 1985). From the point of view of relatedness of antibacterial and hemolytic substances, the 40- and 45-kDa molecules seem to be of particular interest. These proteins are encoded by two distinct genes one of which possesses four different alleles (Roch et al. 1991a). Roch (1979) identified four different protein products corresponding to alleles of the second gene with a pI ranging from 5.9 to 6.3. Celomic fluid of all earthworms contains either two or three isoforms, with one isoform invariably present (pI 6). In European populations of *E. foetida andrei*, three proteins of pI 5.9, 5.95, and 6.3 are encoded by the same gene that expresses three allelic forms. The combination of three alleles (*a*, *b*, and *c*) provides the possibility of six genotypes (*aa*, *bb*, *cc*, *ab*, *ac*, and *bc*) corresponding to six phenotypes (A, B, C, D, E, F). In American populations that originated from Californian ancestors, the fourth allele *d* it present but it has never been detected in the European population. Valembois et al. (1986) explained the occurrence of allele *d*, originally very rare in ancestral European worms, by more favorable conditions for its expression after migration of European ancestors into a new biotope during the 15th century. This opinion can be supported by the fact that even in "Californian" earthworms, the homozygous genotype *dd* has not been proved. The significant antibacterial effect was found mainly in the most frequent phenotype, B (*bb* genotype) and in phenotype K, whose frequency is considered as intermediate. Because of this fact and that bacteriostatic effects of other frequently occurring phenotypes are relatively low, perhaps alternative mechanisms are involved in antibacterial humoral defense.

2.4.2 Lysozyme-Like and Other Antibacterial Substances

As has already been emphasized above, humoral antibacterial defense must be observed in a heat-sensitive antibacterial system associated with hemolytic activity. In 1960 Jollès and Zuili, using the polychaete *Nephthys hombergi*, isolated a lysozyme-like substance that was characterized by relative heat stability at low pH and by antibacterial effects. However, further analyses of molecular weight and amino acid composition revealed that this molecule did not share characteristics of vertebrate lysozyme (Périn and Jollès 1972). Molecules exerting lysozyme-like activity were detected in body fluids and cell lysates of different annelid species (Schubert and Messner 1971; Dales and Dixon 1980). In *Eisenia foetida*, lysozyme-like activity is associated with a 15-kDa protein reacting against both Gram-positive and Gram-negative bacteria and having a pH optimum at 6.2 (Çotukand Dales 1984a, b; Lassalle et al. 1988). Vaccination of earthworms with a fixed bacterial suspension resulted in increased levels of lysozyme-like activity (Lassalle et al. 1988; Hirigoyenberry et al. 1990). Nevertheless, in all the above cases, a structural relatedness with vertebrate lysozyme has not yet been observed. The existence of another antibacterial system was substantiated in a marine polychaete *Glycera dibranchiata* (Anderson 1980; Anderson and Chain 1982; Chain and Anderson 1983a, b). The responsible factor was identified as a 250–450 kDa glycoprotein containing bound divalent cations and at least one disulfide bridge.

2.5 Proteolytic Enzymes in Defense Mechanisms

2.5.1 Proteolytic Enzymes in Invertebrates

In different invertebrate species, proteinase activity has been observed and the role of proteinases in various regulatory pathways documented (Zwilling and Neurath 1981). The mitogenic effects in various systems either under in vivo or in vitro conditions as well as their involvement in meiosis and reproduction have been studied in recent years (Peucellier 1983). One of the most important features of proteinases seems to be the role they play in defense mechanisms.

In insects, crustaceans, and ascidians a complex cascade of serine proteinases and other factors, especially activated by microbial polysaccharides, participate in conversion of the inactive proenzyme, prophenoloxidase to active phenoloxidase (Söderhall and Smith, 1986; Ashida and Yoshida 1988). The prophenoloxidase system, which is responsible for melanization frequently observed in response to invasion of foreign substances or during wound healing, is thought to act as a *non-self* recognition mechanism. It plays a role in host defense not only by mediating encapsulation, but also by providing opsonins, enhancing phagocytosis, and participating in clotting, hemocyte aggregation and the generation of antimicrobial factors (Ratcliffe et al. 1991; Jackson et al. 1993).

2.5.2 Proteolytic Enzymes in Annelids

In annelids, serine proteinases were purified from digestive fluids of the polychaete *Sabellaria alveolata*. All were active at basic pH and their specificity was partially characterized. Despite their origin, they seem to exert many functions, especially those with broad specificities (Peucellier 1983). In oligochaetes, the existence of proteinases with the capacity to digest foreign proteins has been observed in supernatants from celomocyte cultures of *Eisenia foetida* (Valembois et al. 1973). We observed that earthworm celomic fluids of all *Eisenia foetida* and of some *Lumbricus terrestris* efficiently split vertebrate serum proteins (pig IgG–PIgG, human serum albumin–HSA), but not their own celomic fluid proteins nor those of related species (Table 1). More specifically, the proteolytic activity was detected in an approximately 40-kDa fraction separated by gel filtration. Moreover, it was found in inhibition experiments that molecules responsible for proteolytic activities differed from those that mediate hemolytic activities described earlier (Roch 1979). Hemolysis, but not proteolysis was inhibited by simple sugars whereas, in contrast, proteolysis was blocked by 1 mM PMSF that did not influence hemolysis (Tučková et al. 1986a).

Proteinase activity was also found in celomocyte lysates from *Eisenia foetida andrei*. Proteolysis is pH-dependent with two optima at pH 7 and 10, dose-dependent, not influenced by ionic strength, and thermoinsensitive, since heating for 15 min at 100 °C did not completely suppress it. Three serine proteinases, with a molecular weight of about 45 kDa were separated and characterized as trypsin-like and chymotrypsin-like (Roch et al. 1991b).

Table 1. Comparison of the proteolytic effect of *Eisenia foetida* (E.F.) celomic fluid on pig IgG (PIgG) and *Lumbricus terrestris* (L. T.) celomic fluid proteins. (Tučková et al. 1986a)

Sample	Digested protein	cpm in TCA precipitate	TCA precipitable proteins (%)
E.F. celomoc fluid	^{125}I PIgG (10^5 cpm) + PIgG (2 mg/ml)	55 549 ± 365	55.94
PBS		99 285 ± 2397	100.00
E.F. celomic fluid	^{125}I L.T. cel.fluid (10^5 cpm) + L.T. celomic fluid (2 mg/ml)	41 665 ± 320	97.80
PBS		42 595 ± 1067[a]	100.00

Samples were incubated for 24 h at room temperature.
[a]TCA precipitates only about 40% of *L. terrestris* celomic fluid proteins.

We suggested that proteolytic enzymes may be involved not only in nutrition but also in defense mechanisms of annelids. In our later experiments, we measured proteolytic activities in individual samples of celomic fluids of *Eisenia foetida* and *Lumbricus terrestris* and followed the effects after protein stimulation. The pH optimum lies in the alkaline range of 8–10, similar to what was determined for proteinase activity in celomocyte lysates. The bulk of the proteolytic activity at pH 9.5 can be characterized as due to serine proteinases since more than 90% of the activity can be inhibited by 0.5 mM PMSF or 4 μM soybean trypsin (SBT) inhibitor. EDTA or *o*-phenanthroline (1 mM) exerted no inhibitory effect.

The mean specific proteolytic activity determined in celomic fluid of *Eisenia foetida* (1651 ± 227.7 U/mg) was significantly higher than that in *Lumbricus terrestris* (632.0 ± 190.1 U/mg). More individuals with low, or without any detectable proteolytic activity in their celomic fluids under experimental conditions, were found in *L. terrestris* than in *E. foetida*.

2.5.3 Heterogeneity of Proteolytic Enzymes

The heterogeneity of proteolytic activity of individual celomic fluids was demonstrated in SDS gels (7.5 or 10%) with azocasein (1 mg/ml) incorporated into the running gel and cooled during electrophoresis to 0 °C. After washing and incubation at pH 9.5 for 20 min at 37 °C, gels were stained with 0.5% Coomassie Brilliant Blue R-250. Zones of enzymatic activity were indicated by negative staining. SDS-PAGE analyses revealed differences in spectra of azocaseinolytic proteins in two earthworm species. However, differences in specific activity and/or in number of proteolytically active proteins among individual samples were not influenced by primary or secondary stimulation of earthworms with protein antigen—arsanilic acid coupled to human serum albumin (ARS-HSA). However, when the enzymatic activity, estimated independently on protein concentration, was followed we found that the presence of antigen led to increased proteolysis. We have also demonstrated that celomocytes, cultivated in vitro in the presence or absence of antigen, release proteolytic enzymes into the culture medium, and that the increased release (and probably also formation) of enzymes can be induced by protein antigen (Bilej et al. 1993).

2.5.4 Naturally Occurring Proteinase Inhibitors

The earthworms used in experiments were collected in different regions, so their previous stimulation with naturally occurring antigens cannot be excluded. This fact, with respect to enzymatic activity, can explain variations in proteinase activities and SDS-PAGE patterns in individual earthworms. The more frequent occurrence of celomic fluids with undetectable or low proteolytic activities, observed in *L. terrestris* individuals, can be explained by our previous findings showing that *L. terrestris* celomic fluids contain proteinase inhibitors that are able to inhibit the digestion of vertebrate serum proteins by *E. foetida* celomic fluids (Tučková et al. 1986b). Recently two of these inhibitors were successfully isolated (Voburka et al. 1992).

The effect of incubating celomic fluids and cell lysates with 2-mercaptoethanol (2-ME) before SDS-PAGE, which results in an increase of the number of proteolytically active zones, is of particular interest. This effect may be explained in at least two ways: (1) addition of 2-ME results in the reduction of interchain disulfide bonds and leads to separation of polypeptide chains thus increasing the number of proteolytic zones; (2) 2-ME acts as a stimulator of proteolysis and/or causes dissociation of enzyme-inhibitor complexes (Leipner et al. 1993).

3 Adaptive Response to Antigenic Stimulation

Although the ability to discriminate *self* and *non-self* exists in all animal species, our knowledge of the mechanisms that effect recognition is somewhat limited, especially in invertebrates. We can suppose that the substrate specificity of enzymes plays an important role. We have mentioned (Sect. 2.5) that annelids possess a highly active proteolytic system that efficiently digests foreign, but not self proteins (Tučková et al. 1986a). Furthermore, agglutinins involved in invertebrate defense are mostly lectins, so that the interaction of specific sugars with lectins in body fluids and on cell surfaces seems to be functionally important. Moreover, the ability to prevent damage to the host organism by foreign cells, i.e., the ability to reject xenografts, represents another type of specific discrimination of foreign substances. Cytotoxic reactions are accompanied by short-term memory (Cooper 1970). Current information reveals that it is also possible in invertebrates to induce the formation of molecules that possess the ability to interact specifically with immunizing antigen, although the specificity of binding seems to be rather low.

3.1 Adaptive Humoral Response to Antigenic Stimulation

3.1.1 Effect of Antigenic Stimulation on Protein Synthesis

In a recent study we wanted to follow both humoral and cellular responses in *Eisenia foetida* and *Lumbricus terrestris* to parenteral stimulation with a protein antigen. According to our results, the injection of proteins (arsanylated human serum albumin–ARS-HSA, arsanylated bovine gamma globulin–ARS-BGG, human serum albumin–HSA and ferritin) embedded in agar gel, agar gel alone, and sheep erythrocytes into the celomic cavity leads to an increase of total protein concentration in *L. terrestris* and *E. foetida* celomic fluids. The proteins were measured by the modified Bradford method an analyzed by SDS-PAGE. The highest level was determined in *L. terrestris* celomic fluid on day 4 after the secondary dose, when it measured up to three times the average concentration detected in celomic fluids of non-stimulated controls. The increase in total protein concentration in celomic fluids of *E. foetida* was less pronounced. Only a slight increase in celomic fluid proteins was observed after sham stimulation with PBS only. This suggests two points: (1) that the enhanced protein synthesis might be connected with body injury; (2) that the response is relatively specific since control values were less than those of experimental ones.

3.1.2 Formation of Antigen-Binding Protein (ABP)

Celomic fluids from earthworms previously stimulated by protein antigen contain a protein which binds antigen labeled with ^{125}I or with peroxidase (Px), where the antigen had been used for the initial stimulation. The protein also binds to a lesser extent similar proteins to which the earthworms had not been immunized. Thus, specificity of binding appears to be low. The antigen-binding protein (ABP) was undetectable in most non-stimulated earthworms and the slight reaction detectable in a few instances might have been due to some naturally occurring stimulation prior to beginning the experiments. The response was higher when stimulating protein was also used for detection, but in most of the celomic fluids of stimulated worms there was the capacity to bind both ARS-HSA and ARS-BGG regardless of which protein was used for stimulation (Tučková et al. 1988).

Antigen-binding protein was isolated by affinity chromatography using Sepharose 4B with bound ARS-BGG from *L. terrestris* celomic fluid collected on day 8 after secondary stimulation. The molecular mass of ABP was estimated by SDS-PAGE and subsequent immunoblotting in celomic fluid of stimulated *L. terrestris*. As shown in Fig. 2Aa the protein band reacting with Px-ARS-HSA after transfer to nitrocellulose had a mol. wt. of about 56 kDa. Analyses of affinity-purified ABP on SDS-PAGE under non-reducing conditions revealed the presence of one major band, again of mol. wt. 56 kDa. After reduction with 2-mercaptoethanol (2-ME) the original 56-kDa band disappeared and was re-

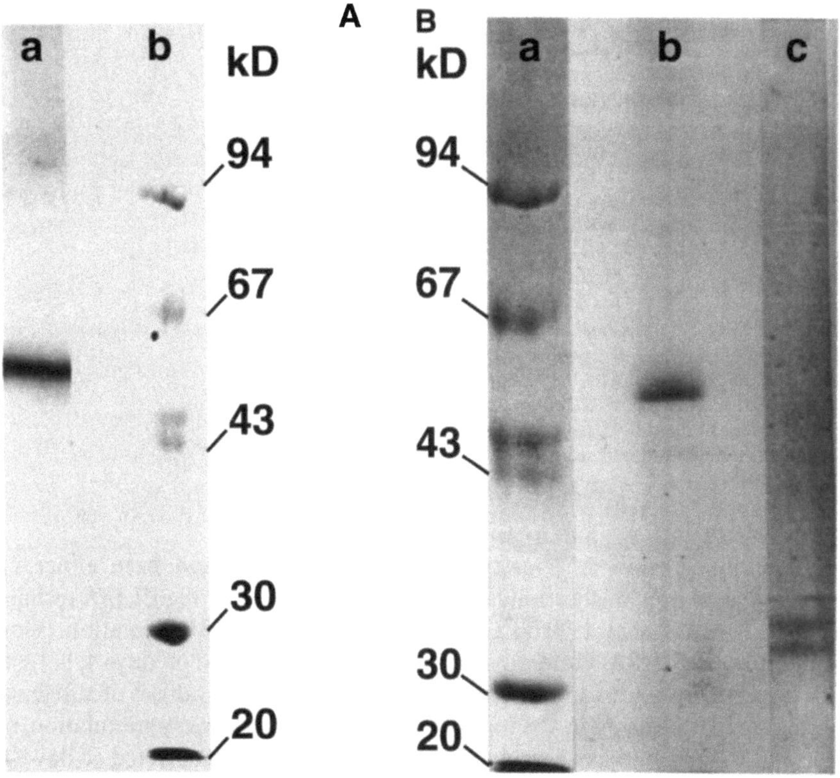

Fig. 2. **A** Western blot analysis of celomic fluid obtained from *L. terrestris* immunized with ARS-HSA: *a* reaction with Px-ARS-HSA; *b* molecular weight markers (Pharmacia, Uppsala) stained with Coomassie Brilliant Blue R-250. **B** SDS-PAGE of affinity chromatography-purified ABP: *a* molecular mass markers. The sample was subjected to electrophoresis on 10% acrylamide slab gel under *b* non-reducing and *c* reducing conditions (heated at 100 °C for 3 min in 1% SDS in the presence of 2% 2-mercaptoethanol). (Tučková et al. 1991a)

placed by two 31- and 33-kDa bands (Fig. 2Bc), suggesting that the APB molecule is composed of two disulfide-linked polypeptide chains. ABP, after isolation, retained its binding activity (Table 2), while after reduction with 2-ME, the two separate chains did not reveal detectable binding capacity. This indicates that both polypeptide chains are involved in antigen-binding site formation (Tučková et al. 1991a).

3.1.3 Monoclonal Antibodies to Antigen-Binding Protein

A panel of monoclonal antibodies (mAbs) against the ABP molecule has been prepared and three of them 6D4/2, 6D4/3 and 5G9/3 (all IgG1, kappa isotype) were selected and tested using the ELISA technique for reactivity with celomic

Table 2. The binding of ^{125}I-ARS-HSA and ^{125}I-ARS-BGG by celomic fluid fractions obtained by affinity chromatography on ARS-BGG-Sepharose as determined by dot-blot assay. (Tučková et al. 1991a)

Fraction	Binding of ^{125}I labelled (%)	
	ARS-BGG	ARS-HSA
Unadsorbed	16.6	29.4
Eluate I	8.3	9.6
Eluate II	60.9	40.8

Elution was performed with 0.3 M borate buffer, pH 8 (eluate I) and 0.018 M glycine-HCl buffer, pH 2.4. (eluate II) The values were obtained by densitometric analyses of autoradiograms on a Beckman densitometer R 112, integrator R 115.

fluid obtained from *L. terrestris* and *E. foetida* that had been either nonstimulated or previously stimulated with ARS-HSA using the ELISA technique (Fig. 3; Tučková et al. 1991b). To follow the kinetics of ABP formation, celomic fluid from ARS-HSA-stimulated earthworms was collected on days 4, 8, 12, and 19, respectively, after administering primary and secondary doses of antigen and tested by ELISA using mAbs to ABP. Four days after primary stimulation, only a low mAb binding was observed. The highest reaction was detected on day 8 and a decline of the reactivity was observed on day 19. After the secondary challenge, the highest antigen-binding capacity was measured between days 4 and 6 and was significantly higher than that found after the primary antigenic challenge. Similar results were obtained using all three mAbs (Table 3). The time course of ABP responses was similar in both species, but the actual values were lower in *E. foetida* compared to those obtained in *L. terrestris* celomic fluids. The lower binding ability of *E. foetida* celomic fluids might be connected with the high proteolytic activity demonstrated in these fluids which are able to efficiently split the antigen (Tučková et al. 1986a; Leipner et al. 1993). The kinetics of response to ARS-HSA detected with mAbs to ABP corresponded to that detected with Px- or ^{125}I-labeled antigen (Tučková et al. 1988), therefore one can assume that mAbs react with ABP of earthworms and that the ABP molecules of both earthworm species are similar, if not identical.

Furthermore, mAb 6D4/2 was used in ABP separation, Western blot analyses, and immunoprecipitation of ARS-HSA stimulated celomic fluids. As in further experiments, the only celomic fluid protein reacting with mAb was the 56-kDa molecule. When the complex mAb-ABP was dissolved and separated on SDS-PAGE, and when the protein bands were transferred onto nitrocellulose sheets, again only the 56-kDa band bound the labeled antigen (Px-ARS-HSA). Moreover, the reactivity of ABP with antigen and mAb, together with our knowledge of the existence of only one binding site per ABP, may indicate that

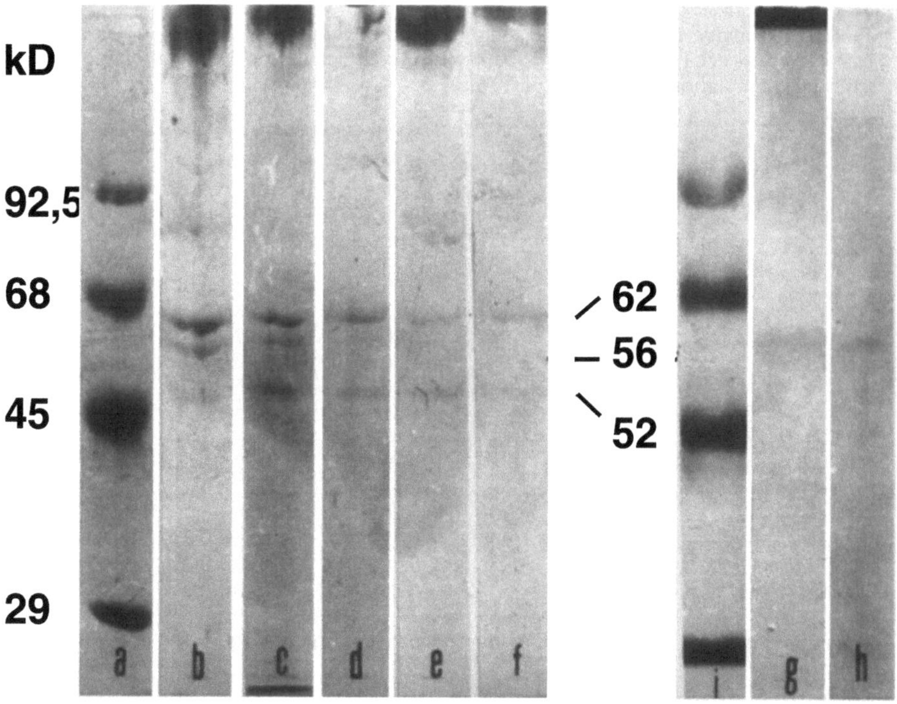

Fig. 3. SDS-PAGE and Western blot analyses of immunoprecipitates formed after interaction of 6D4/2 (*b*) and 5G9/3 (*c*) mAb and ovine anti-mouse Ig antibodies with the celomic fluid of ARS-HSA-stimulated earthworms *L. terrestris*. The controls where mAb was either omitted (*d*), replaced with Balb/c peritoneal washing (*e*) or where normal Balb/c serum instead of celomic fluid was used (*f*), showed that only 56-kDa celomic fluid protein reacted with the mAb. The Western blot analyses of the celomic fluid (*g*) and of the precipitate (*h*), SDS-PAGE analysis (*b*), confirmed that only 56-kDa protein bound Px-labeled ARS-HSA. Lanes *a* and *i* represent molecular mass standards. (Tučková et al. 1991b)

the mAb specific for the ABP molecule seems to react with an antigenic determinant located outside the ABP binding site.

The specificity of mAb was also confirmed in experiments where reactions with celomic fluid were inhibited (by up to 60%), in a dose-dependent manner with the immunizing antigen (ARS-HSA) and also with a similar protein (Tučková et al. 1991b). The low specificity of the ABP-binding site already mentioned (Sect. 2.1.2) may reflect a low ability of earthworms to discriminate between foreign antigens. This is also apparent from the data published by Cooper (1970) and Hildemann (1981), who showed that earthworms displayed weak rejection responses to allografts, whereas they readily rejected xenografts. The reaction to first grafts certainly seems to determine the fate of the second-set response. The memory response to second allografts is short-lived, no more than 10 days, while the response to second xenografts varies from short- to long-lived (Cooper and Roch 1986).

Table 3. Reactivity of mAb (monoclonal antibody) with celomic fluids of earthworms *Lumbricus terrestris* (L.T.) and *Eisenia foetida* (E.F.) collected during the course of primary and secondary stimulation. (Tučková et al. 1991b)

				ELISA (O.D. ± SD) mAb		
Animal	Antigen	Dose	Day	6D4/2	6D4/3	5G9/3
L.T.	ARS-HSA	I	4	0.15 ± 0.03	0.21 ± 0.02	0.23 ± 0.01
			8	0.45 ± 0.14	0.47 ± 0.02	0.46 ± 0.08
			12	0.23 ± 0.02	0.28 ± 0.01	0.23 ± 0.03
		II	4	0.46 ± 0.03	0.47 ± 0.02	0.47 ± 0.04
			8	1.80 ± 0.05	1.87 ± 0.08	1.62 ± 0.02
			12	1.16 ± 0.34	1.55 ± 0.24	1.24 ± 0.05
	O[a]			0.23 ± 0.08	0.25 ± 0.06	0.31 ± 0.10
E.F.	ARS-HSA	I	4	0.30 ± 0.13	0.29 ± 0.14	0.28 ± 0.19
			8	0.41 ± 0.07	0.37 ± 0.02	0.36 ± 0.10
			19	0.26 ± 0.08	0.34 ± 0.04	0.30 ± 0.02
		II	4	0.49 ± 0.01	0.56 ± 0.03	0.50 ± 0.02
			8	0.46 ± 0.04	0.33 ± 0.03	0.33 ± 0.05
	O[a]			0.24 ± 0.12	0.30 ± 0.07	0.27 ± 0.09

Values are expressed as means ± SE of four independent experiments.
[a]Controls are means of 10 estimations.

3.2 Cellular Adaptive Response to Antigenic Stimulation

3.2.1 Cellular Expression of Antigen-Binding Protein

Following the previous work, the presence of ABP molecules on surfaces of free celomocytes harvested from antigen-stimulated earthworms has also been observed. The superficial binding of the antigen was detected either by direct measurement of radioactivity and microautoradiography (Bilej et al. 1990b) or by flow cytometric analysis (Bilej et al. 1991c). The time course of antigen-binding capacity of free celomocytes agrees with that which revealed the occurrence of ABP in celomic fluid: The maximal percentage of positive cells as well as the maximal amount of antigen bound per single cell were observed on day 8 after in vivo stimulation. Comparisons of the increase in numbers of labeled cells and in amounts of bound antigen revealed that the concentration of superficial binding sites must be enhanced significantly. Whereas the number of antigen-binding celomocytes reaches twice the value of non-stimulated controls on the 8th day, the level of totally bound antigen is augmented almost three times.

Inhibition experiments confirmed the relatively low specificity of ABP, which failed to reach values reported by Laulan et al. (1985). The binding of labeled antigen used for in vivo stimulation can be inhibited by preincubation of cell suspensions with non-labeled antigen or similar proteins; however, the highest inhibiting effect was observed when the immunizing antigen was used. It should be noted that based upon flow cytometric analysis, high degrees of non-specific (or basal) antigen binding were detected—up to 48% of cells—but these cells possessed low concentrations of antigen-binding sites. After stimulation, the number of positive cells increased up to 60% but the expression of ABP was enhanced more than two times. From this point of view the inhibition of antigen binding with cold antigen reached almost 100%, whereas using other similar cold proteins produced 73–83% (after subtraction of basal binding).

3.2.2 Binding of Monoclonal Antibodies

Similar results were obtained when the binding of monoclonal antibodies against ABP was followed (Tučková et al. 1991b). Non-specific binding (26% of cells) may reflect in part, the presence of IgG-binding molecules on cell surfaces (Rejnek et al. 1986). When the cells collected from ARS-HSA-stimulated earthworms were analyzed by flow cytometry the small peak of highly positive cells occurred which had never been detected previously in control samples and which disappeared after preincubation of cells with cold antigen or another similar protein. The preincubation of cells with ARS-HSA (antigen used for stimulation) or with HSA led to the decrease of monoclonal antibody binding. Though the monoclonal antibodies reacted with antigenic determinant located outside the binding site of ABP, the inhibition may be explained as having been due to a steric hindrance (Fig. 4).

3.2.3 Cells Involved in Antigen Recognition and Binding and the Role of the Mesenchymal Lining

The presence of highly positive cells occurring after antigenic stimulation indicates the involvement of "specialized" cells. One of the possible candidates could be agranular cells corresponding to neutrophilic celomocytes (Cooper and Stein 1981); these cells are known to constitute one of the effector cell-types in the rejection of xenogeneic grafts. In our experiments these cells react with gold-labeled antigen to internalize it (Fig. 5; Bilej et al. 1991c). However, these results do not solve the question of whether these cells originate directly from the mesenchymal lining or if they differentiate and proliferate from free celomocytes. Due to this lack of information we tried to follow the antigen-induced proliferation of free celomocytes (Bilej et al. 1992a). Celomocytes collected at various time intervals after parenteral stimulation were cultivated in the presence of ^{3}H-thymidine. Surprisingly, the level of ^{3}H-thymidine incorporation decreased and reached its lowest value on day 8 when both humoral and cellular binding activities were maximal. After a second injection of antigen on day 16, the proliferative response rose rapidly. Similarly, if the cells collected from control unimmunized earthworms were stimulated with antigen in vitro, proliferation reached only 50% of the levels of unstimulated controls and was comparable to that of cells from antigen-stimulated worms. In contrast, the presence of antigen

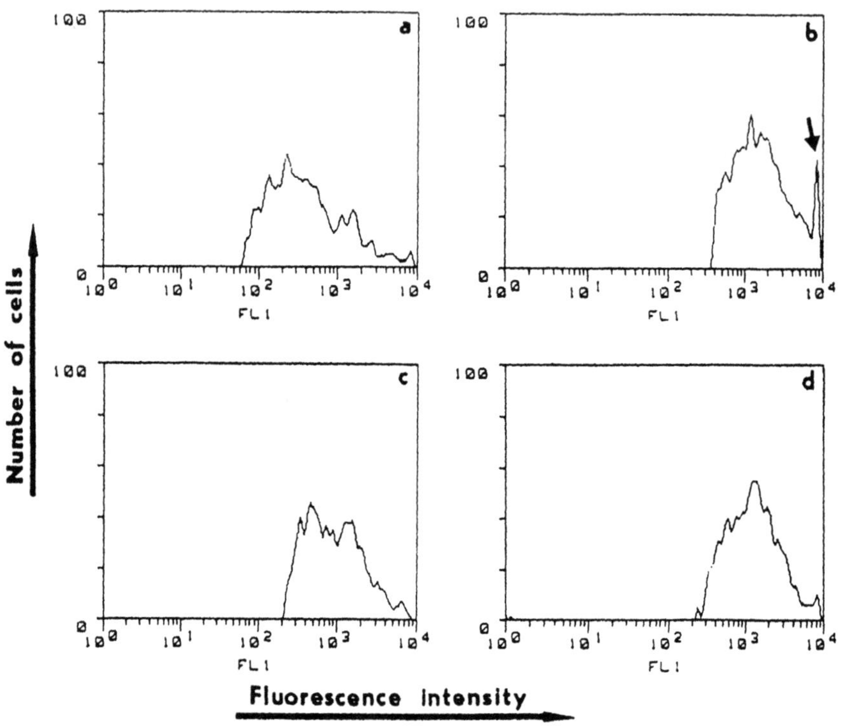

Fig. 4. Flow cytometric analysis of mAb binding on celomocyte surface. **a** Celomocytes isolated from non-stimulated animals incubated with mAb. **b** Celomocytes isolated from ARS-HSA-stimulated animals incubated with mAb. **c** Celomocytes isolated from ARS-HSA-stimulated animals incubated with ARS-HSA before application of mAb. **d** Celomocytes isolated from ARS-HSA-stimulated animals incubated with HSA before application of mAb. Representative data of one of the five independent experiments. *Arrow* indicates peak of cells with high density of binding sites which occurs in non-inhibited samples from stimulated animals only. (Tučková et al. 1991b)

in cultures of cells from worms pre-stimulated in vivo resulted in significantly augmented proliferation of cells. If another (but similar) protein was used for the second contact in vitro the response reached an intermediate level. Precursor cells in the mesenchymal lining of the celomic cavity responded to stimulation with protein antigen in vivo by proliferation almost immediately. Thus antigenic stimuli may trigger proliferation and differentiation of precursor cells in the mesenchymal lining. The cells then enter the celomic cavity and after repeated contact with the same antigen these "pre-stimulated" cells undergo further mitotic cycles.

A key role of the mesenchymal lining in antigen-induced adaptive responses has also been observed when the fate of parenterally administered ^{125}I-labeled antigen is followed (Rejnek et al. 1993). Within the first 24 h after administration of labeled antigen, radioactivity was observed in all tissues except the epidermis and cuticle. Later, on day 4, the label above muscle cells disappeared and

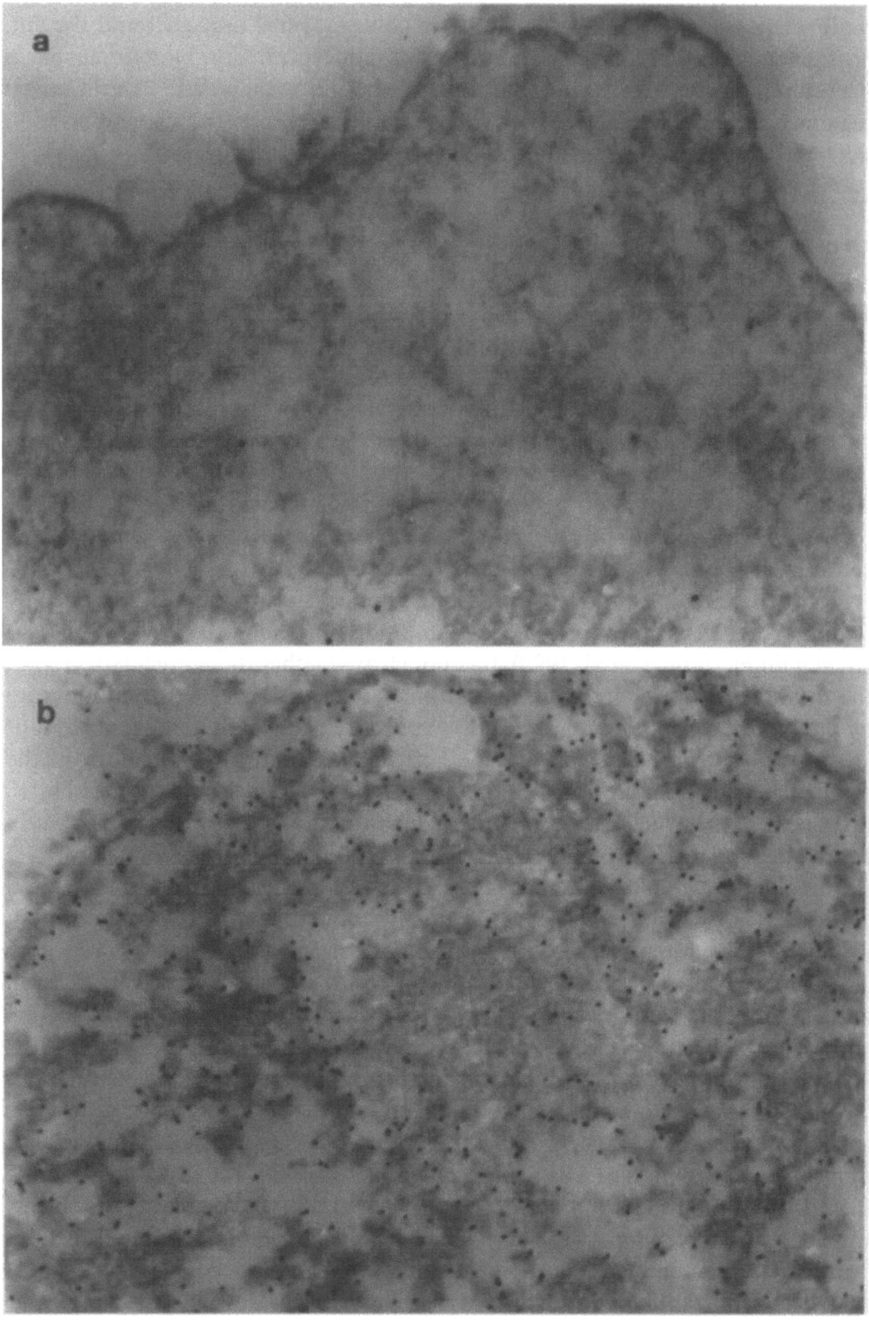

Fig. 5. Detection of gold-labeled antigen in the celomocytes of *E. foetida*. **a** Sparsely disseminated gold particles in the cytoplasm of translucent agranular (probably neutrophilic) celomocytes following an in vitro incubation of non-presensitized cells. **b** Plentiful intracytoplasmic gold particles in the agranular translucent cell following the antigen challenge of a presensitized cell sample. (Bilej et al. 1991c)

radioactivity remained detectable only in chloragogenic tissue around the gut, dorsal vein, and typhlosole. The antigen was concentrated in interstitial spaces between chloragogen cells and was frequently bound apically. Whether the apical regions represent predilective sites with possible receptor structures involved in *self/non-self* recognition remains to be solved.

3.2.4 Antigen Processing in Earthworms

The recognition of antigen by helper T cells of mammals requires antigen processing and antigen presentation. Antigen is internalized and degraded by proteolytic enzymes, and peptide fragments are displayed on the surface of antigen-presenting cells in association with MHC class II molecules. Though we have not found and perhaps should not expect such sophisticated mechanisms in invertebrates where lymphocytes and antibody formation have not been observed, we can at least attempt to compare the mechanisms that lead to antibody responses with the adaptive responses described in earthworms emphasizing certain general analogies.

Protein antigen administered into the celomic cavity is rapidly degraded by strong proteolytic enzymes (see Sect. 2.5) and is bound by cells of the mesenchymal lining (Rejnek et al. 1993). It is not clear whether these precursor cells in the mesenchymal lining recognize and bind intact antigen or its peptide fragments, but the second possibility seems to be more plausible and more important (see below). Nevertheless, proteolytic activity has been observed both in the celomic fluid and in free celomocytes (Bilej et al. 1993; Rejnek et al. 1993). In brief, up to 80% of the injected antigen is degraded proteolytically during the first 48 h. Moreover, when the proteolytic activity of supernatants from free celomocyte cultures was tested, the higher level of activity was detected in antigen-stimulated cultures. This finding suggests an inducible component that is involved in proteinase formation (Bilej et al. 1993).

The rapid digestion of protein antigen contrasts with a relatively delayed antigen-binding protein response which reaches maximal values on day 8. This contrast has been partially solved by in vitro experiments in which ABP formation in tissue cultures of small explants of gut wall with adjacent mesenchymal lining was followed (Tučková and Bilej 1994). The presence of ABP in tissue culture supernatants was demonstrated by the ELISA technique with monoclonal antibodies to ABP with similar kinetics compared to ABP of celomic fluid. By scanning electron microscopy (Bilej et al. 1994) two main types of cells were distinguished which were derived from tissue explants during cultivation. The first type resembled chloragogen cells, whereas the second was a round-shaped less differentiated cell. These same cell types were described by Janda and Bohuslav (1934) in their 60 year old paper, thus these results confirm the plausibility of our hypothesis (Bilej et al. 1992a) that an administered antigen is first recognized by precursor cells in the mesenchymal lining; these cells proliferate, differentiate, and then enter the celomic cavity to fulfill their defense function.

The effect of proteolytic degradation has been studied in experiments cultivating tissue explants in the absence or in the presence of a non-toxic inhibitor of serine proteinases (Pefabloc; 4-(2-aminoethyl)-benzenesulfonyl fluoride; Boehringer). In antigen-stimulated tissue cultures, ABP formation was observed with maximal levels reached on day 5. When the culture medium was supplemented with a proteinase inhibitor the ABP response was almost absent. In contrast, the efficient signal that triggers ABP formation was supplied by proteolytic fragments (< 20 kDa) of antigen both in the presence and absence of the inhibitor. Fragments of a mol. wt. greater than 20 kDa caused no stimulation, while the small fragments (< 3 kDa) were the most effective. The question is why the non-stimulating, large fragments are not split further during cultivation. The only plausible current explanation we can consider is that the binding of antigen to superficial ABP molecules does not always provide the proper signal that triggers ABP formation but that the activation signal can be realized only by peptides of certain sizes and/or structural motifs. In any case, we suggest that proteolytic processing is involved in cell stimulation that leads to ABP formation.

4 Celomocyte Superficial Molecules

The occurrence of antigen-binding protein on celomocyte surfaces is certainly associated with adaptive type responses. Nevertheless, there are other superficial molecules that are expressed constitutively and are assumed to play a receptor role.

4.1 Agglutinins

Earthworm celomocytes can form rosettes with vertebrate erythrocytes (Cooper 1973; Kauschke and Mohrig 1987) probably as a result of membrane bound or secretory agglutinin prior to release in the celomic fluid. The hemagglutinating activity of the celomic fluid was inhibited by sugars (Stein and Cooper 1983) indicating that active molecules can have a lectin-like character. Celomocytes of *L. terrestris* are involved both in E-type rosette formation (single layer of erythrocytes adhering to cell surface) and in the secretory type of rosettes (multiple layer of erythrocytes). Due to the fact that cyclohexamide treatment does not affect rosette formation, presumably hemagglutinins are stored intracellularly prior to secretion. Furthermore, celomocytes display allotypic specificity by forming rosettes with erythrocytes from different individuals of the same species (Stein and Cooper 1988). Though the agglutinins are expressed constitutively their inducible character has also been observed (Stein et al. 1982; Wojdani et al. 1982). Hemagglutinin levels increased considerably after a single injection of erythrocytes or carbohydrates. Elevated agglutinin titers did not correlate directly with non-specific increase of total protein concentration. Moreover, one or two new proteins have been detected (Stein et al. 1990). An effort to determine the biological importance of agglutinins in inactivating microbes has led to the observation of bacterial agglutinins (Stein et al. 1986).

4.2 Receptors for Mitogens

Experiments testing the proliferative responses of celomocytes after mitogen stimulation revealed other interesting results. Roch et al. (1975) observed that all celomocytes are able to bind concanavalin A but only a small subset (less than 1%) responds by ^{3}H-thymidine incorporation. Augmented incorporation of ^{3}H-thymidine was detected after phytohemagglutinin treatment in non-adherent celomocytes of *L. terrestris* (Toupin and Lamoureux 1976) and in large adherent cells of *E. foetida* (Roch 1977).

4.3 β_2-Microglobulin-Like Molecule

Due to the generally accepted observation that invertebrates do not possess either immunoglobulins or lymphocytes, the search for immunoglobulins seems to be useless. Nevertheless, there are certain molecules displaying considerable homology with vertebrate immunoglobulin superfamily members, and molecules reacting with such vertebrate proteins. For example, Shalev et al. (1980, 1981) demonstrated the cross-reactivity of β_2-microglobulin sera in total extracts of several invertebrates including earthworms. Two years later, Roch and coworkers (Roch and Cooper 1983; Roch et al. 1983) reported the presence of a β_2-microglobulin-like determinant on the membrane of a celomocyte subpopulation identified as weakly adherent acidophils. Inhibition and competition experiments indicate that the homology between the β_2-microglobulin-like molecule in earthworms and human β_2-microglobulin can be restricted to only one major determinant. The existence of a β_2-microglobulin-like molecule in earthworms may provide important information about the evolution of β_2-microglobulin and its role in *self/non-self* recognition, since it forms the light chain of the class I MHC molecule.

4.4 IgG-Binding Protein

We have tested a number of antisera to vertebrate Ig isotypes and their subunits for cross-reactivity with *Lumbricus terrestris* and *Eisenia foetida* celomic fluids. We have shown that sheep and goat fast-migrating IgG molecules and their $F(ab)_2$ fragments react with *L. terrestris* and *E. foetida* celomic fluid proteins. The reaction was not mediated by the antibody binding site, but by some other component of the IgG molecule, since affinity purified antibodies of different antigen specificities as well as IgG from normal sheep sera reacted with the same celomic fluid proteins. The presence of at least two reacting proteins of mol. wts. 47 kDa and 50 kDa in *L. terrestris* and of mol. wt. 42 kDa in *E. foetida* was observed. Sheep antibodies and IgG molecules also bind to the surface of about 20% of celomocytes and immunoprecipitation analyses showed that the IgG-binding proteins in celomic fluids and in cell lysates were similar (Rejnek et al. 1986).

4.5 SpA-Binding Protein

Immunoglobulins are the only serum proteins of vertebrates that react with Staphylococcal protein A (SpA; Goudswaard et al. 1978; Marchalonis et al. 1978; Zikán et al. 1980; Langone 1982; Richman et al. 1982). Even though immunoglobulins have never been found in earthworms, nor in any other invertebrate species, they respond to parenteral administration of various antigens by formation of the celomic fluid protein that binds the administered antigen—ABP. Therefore, we considered it interesting to test whether this or some other celomic fluid protein reacts with SpA. Surprisingly, a protein reacting with SpA was found in the celomic fluids of earthworms, both *L. terrestris* and *E. foetida*, and the level of this protein increased in worms stimulated with protein antigens. The binding experiments (using ^{125}I-SpA and ^{125}I-ARS-HSA) demonstrated that ABP reported earlier (Tučková et al. 1988, 1991a) is different from that responsible for binding SpA (Table 4).

The SpA-binding protein was isolated from celomic fluids of *L. terrestris* and *E. foetida* by affinity chromatography on SpA-Sepharose. SDS-PAGE analyses,

Table 4. The binding of ^{125}I-labelled SpA and ARS-HSA by celomic fluid fractions from non-stimulated and ARS-HSA-stimulated *L. terrestris* earthworms. (Rejnek et al. 1991)

Fraction	Protein administered	Binding of ^{125}I-SpA (cpm)	Binding of ^{125}I-ARS-HSA (cpm)
Original CF	0	6991.2 ± 1314.0	419.0 ± 136.7
Unadsorbed	0	1202.2 ± 319.4	412.0 ± 87.4
Eluate I	0	2022.1 ± 284.9	424.6 ± 116.8
Eluate II	0	2601.0 ± 356.3	369.6 ± 34.3
Original CF	ARS-HSA	8568.1 ± 1791.3	1415.4 ± 267.1
Unadsorbed	ARS-HSA	1396.7 ± 471.5	1071.7 ± 102.3
Eluate I	ARS-HSA	1555.1 ± 121.7	412.3 ± 66.2
Eluate II	ARS-HSA	3206.6 ± 207.5	415.0 ± 80.1

Adsorbed material on immunoadsorbent was eluted with 0.3 M borate buffer, pH 8 (eluate I) and 0.018 M glycine-HCl buffer, pH 2.4 (eluate II). Dots on nitrocellulose were cut out and radioactivity measured and expressed in counts per minute (cpm). Values are mean ± SE of three independent experiments.

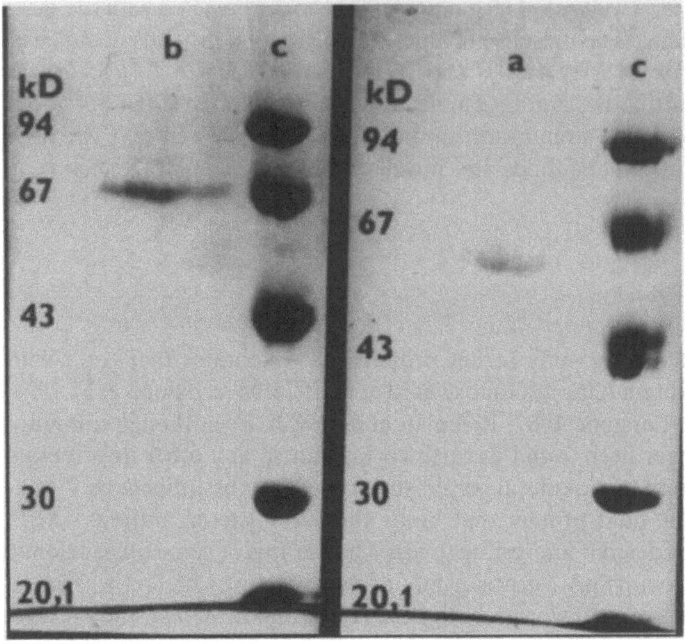

Fig. 6. SDS-PAGE analyses of isolated SpA-binding protein (eluate II) under *a* non-reducing and *b* reducing conditions. *c* Molecular weight markers (Pharmacia, Uppsala); stained with Coomassie blue R-250. (Rejnek et al. 1991)

under reducing and non-reducing conditions, showed that SpA-binding activity was associated with a single chain protein, with mol. wt. 62 kDa (Fig. 6). The carbohydrate moiety of this protein consists of 21.4 g/100 g of O-linked and N-linked oligosaccharides. The amino acid composition of SpA-binding protein showed no structural homology with that of human IgG1 heavy chain (Daw; SΔQ = 169) which also binds SpA (Rejnek et al. 1991). The flow cytometric analyses of free celomocytes revealed that 50% of *L. terrestris* celomocytes reacted with SpA in vitro, whereas in *E. foetida* the number of positive cells did not exceed 10% (Bilej et al. 1992b).

Takahashi et al. (1992) have recently discovered the immunoglobulin joining (J) chain in the earthworm genome. Though the structural homology of some invertebrate proteins and their vertebrate counterparts has been substantiated, it is difficult to decide whether these proteins play a defense function in the absence of simultaneous functional assays. For example, hemolin is involved in invertebrate defense and is homologous with members of the immunoglobulin superfamily (Sun et al. 1990). Nevertheless, a continuous search for invertebrate molecules homologous to proteins of the immunoglobulin superfamily may clarify considerably the evolution of *self*/*non-self* recognition.

5 Concluding Remarks

In conclusion, we can question whether there are receptors in invertebrates; perhaps a positive answer is possible. For example, the existence of receptor structures based on *self/non-self* recognition was substantiated in transplantation experiments (recently summarized by Cooper and Roch in Větvička et al. 1994), in cell-mediated cytotoxicity (Valembois et al. 1980) and in the production of antigen-binding proteins etc. There is no doubt that earthworms recognize *self* and *non-self* but the ability to discriminate between *non-self* is considerably less evolved when compared to vertebrates. In line with this view, Janeway (1992) proposed that receptors providing broad-spectrum recognition of harmful foreign materials were the ancestral recognition structures. These non-clonal receptors (pattern based receptors) appeared to detect common constituents of pathogenic microorganisms. The modification of such microbial structures requires changing these common constituents and can lead even to the loss of pathogenicity. Janeway's opinion is confirmed by experimental data indicating the presence of SpA-binding protein (see Sect. 4.5) or receptors for lipopolysaccharides (Roch 1977) on the celomocyte surface. We can conclude that the main defense mechanisms in earthworms are represented by the innate response (phagocytosis, humoral factors), but the role of adaptive response based on the existence of polyspecific broad-spectrum recognition of *non-self* cannot be overlooked.

References

Alonso-Bedate M, Sequeros E (1985) Suggested regulatory mechanisms for caudal regeneration in *Allolobophora molleri* (Annelida; Oligochaeta). Comp Biochem Physiol 81A: 225–228

Anderson RS (1980) Antibacterial activity in the coelomic fluid of a polychaete annelid, *Glycera dibranchiata*. Biol Bull 159: 259–268

Anderson RS, Chain BM (1982) Antibacterial activity in the coelomic fluid of a marine annelid *Glycera dibranchiata*. J Invertebr Pathol 40: 320–326

Ando K, Natori S (1988) Molecular cloning, sequencing, and characterization of cDNA for Sarcotoxin IIA, an inducible antibacterial protein of *Sarcophaga peregrina* (flesh fly). Biochemistry 27: 1715–1721

Ashida M, Yoshida H (1988) Limited proteolysis of prophenoloxidase during activation by microbial products in insect plasma and effect of phenoloxidase on electrophoretic mobilities of plasma proteins. Insect Biochem 18: 11–19

Baba K, Okada M, Kawano T, Komano H, Natori S (1987) Purification of Sarcotoxin III, a new antibacterial protein of *Sarcophaga peregrina*. J Biochem 102: 69–74

Banerjee A, Datta PK, Basu PS, Datta TK (1991) Characterization of a naturally occurring protease inhibitor in the hemolymph of the scorpion, *Heterometrus bengalensis*. Dev Comp Immunol 15: 213–218

Bang FB (1973) A survey of phagocytosis as a protective mechanism against disease among invertebrates. In: Braun W, Unger J (eds) Non-specific factors influencing host resistance. Karger, Basel, pp 2–10

Bilej M, Větvička V, Tučková L, Trebichavský I, Koukal M, Šíma P (1990a) Phagocytosis of synthetic particles in earthworms. Effect of antigenic stimulation and opsonization. Folia Biol (Prague) 36: 273–280

Bilej M, Tučková L, Rejnek J, Větvička V (1990b) In vitro antigen-binding properties of coelomocytes of *Eisenia foetida* (Annelida). Immunol Lett 26: 183–188
Bilej M, Scheerlinck JP, VandenDriessche T, De Baetselier P, Větvička V (1991a) The flow cytometric analysis of in vitro phagocytic activity of earthworm coelomocytes (*Eisenia foetida*; Annelida). Cell Biol Int Rep 14: 831–837
Bilej M, De Baetselier P, Trebichavský I, Větvička V (1991b) Phagocytosis of synthetic particles in earthworms: absence of oxidative burst and possible role of lytic enzymes. Folia Biol (Prague) 37: 227–233
Bilej M, Rossmann P, VandenDriessche T, Scheerlinck JP, De Baetselier P, Tučková L, Větvička V, Rejnek J (1991c) Detection of antigen in the coelomocytes of the earthworm *Eisenia foetida* (Annelida). Immunol Lett 29: 241–246
Bilej M, Šíma P, Slipka J (1992a) Repeated antigenic challenge induces earthworm celomocyte proliferation. Immunol Lett 32: 181–184
Bilej M, Rejnek J, Tučková L (1992b) The interaction of staphylococcal protein A with free coelomocytes of annelids. Cell Biol Int Rep 16: 481–485
Bilej M, Tučková L, Rejnek J (1993) The fate of protein antigen in earthworms: study in vitro. Immunol Lett 35: 1–6
Bilej M, Tučková L, Rossmann P (1994) A new approach to in vitro studies of antigenic response in earthworms. Dev Comp Immunol 18: 363–367
Boman HG, Steiner H (1981) Humoral immunity in cecropia pupae. Curr Top Microbiol Immunol 94/95: 75–89
Boman HG, Faye I, v. Hofstein P, Kockum K, Lee JY, Xanthopoulos KG (1985) On the primary structures of lysozyme, cecropins and attacins from *Hyalophora cecropia*. Dev Comp Immunol 9: 551–558
Cameron GR (1932) Inflammation in earthworms. J Pathol 35: 933–972
Casteels P, Ampe C, Jacobs F, Vaeck M, Tempst P (1989) Apidaecins: antibacterial peptides from honeybees. EMBO J 8: 2387–2391
Chain BM, Anderson RS (1983a) Antibacterial activity of the coelomic fluid of the polychaete, *Glycera dibranchiata*. I. The kinetics of the bactericidal reaction: Biol Bull 164: 28–40
Chain BM, Anderson RS (1983b) Antibacterial activity of the coelomic fluid of the polychaete, *Glycera dibranchiata*. II. Partial purification and biochemical characterization of the active factor. Biol Bull 164: 41–49
Cooper EL (1965) Rejection of body-wall xenograft exchanged between *Lumbricus terrestris* and *Eisenia foetida* Am Zool 5: 665
Cooper EL (1970) Transplantation immunity in helminths and annelids. Transplant Proc 2: 216–221
Cooper EL (1973) Evolution of cellular immunity. In: Braun W, Unger J (eds) Non-specific factors influencing host resistance. Karger, Basel, pp 11–23
Cooper EL, Roch P (1986) Second-set allograft responses in the earthworm *Lumbricus terrestris*. Kinetics and characteristics. Transplantation 41: 514–520
Cooper EL, Stein EA (1981) Oligochaetes. In: Ratcliffe NA, Rowley AF (eds) Invertebrate blood cells, vol 1. Academic Press, London, pp 75–140
Cooper EL, Acton RT, Weinheimer PF, Evans EE (1969) Lack of bacteriocidal response in the earthworm *Lumbricus terrestris* after immunization wtih bacterial antigens. J Invertebr Pathol 14: 402–406
Çotuk A, Dales RP (1984a) The effect of the coelomic fluid of the earthworm *Eisenia foetida* Sav. on certain bacteria and the role of the coelomocytes in the internal defence. Comp Biochem Physiol 78A: 271–275
Çotuk A, Dales RP (1984b) Lysozyme activity in the coelomic fluid and coelomocytes of the earthworm *Eisenia foetida* Sav. in relation to bacterial infection. Comp Biochem Physiol 78A: 469–474
Dales RP, Dixon LJR (1980) Responses of polychaete annelids to bacterial infection. Comp Biochem Physiol 67A: 391–396
Dales RP, Kalaç Y (1992) Phagocytic defence by the earthworm *Eisenia foetida* against certain pathogenic bacteria. Comp Biochem Physiol 101A: 487–490

Faulhaber LM, Karp RD (1991) A diphasic immune response against injected bacteria in the American cockroach. Dev Comp Immunol (Suppl) 1: 47
Golding DW (1974) Regeneration and growth control in *Nereis*. III. Separation of wound healing and segment regeneration by experimental endocrine manipulation. J Embryol Exp Morphol 32: 99–109
Götz P, Trenczek T (1991) Antibacterial proteins in insects other than *Lepidoptera* and *Diptera* and some other invertebrates. In: Gupta AP (ed) Immunology of insects and other arthropods. CRC Press, Boca Raton. pp 323–346
Goudswaard J, van der Dponk JA, Noordzig A, van Dam RH, Vaerman JP (1978) Protein A reactivity of various mammalian immunoglobulins. Scand J Immunol 8: 21–28
Herlant-Meewis H (1966) Les cellules neurosecretrices de la chaine nerveuse d'*Eisenia foetida*. Z Zellforsch 69: 319–325
Herlant-Meewis H, Deligne J (1964) Regeneration in annelids. In: Abercrombie M, Brachet J (eds) Advances in morphogenesis, vol 4. Academic Press, New York, pp 155–215
Hildemann WH (1981) Immunophylogeny: from sponges, to hagfish to mice. In: Hildemann WH (ed) Frontiers in immunogenetics. Elsevier, Amsterdam, pp 3–19
Hirigoyenberry F, Lassalle F, Lassègues M (1990) Antibacterial activity of *Eisenia fetida andrei* coelomic fluid: transcription and translation regulation of lysozyme and proteins evidenced after bacterial infestation. Comp Biochem Physiol 95B: 71–75
Hrženjak T, Hrženjak M, Kašuba V, Efenberger-Marinculič P, Levanat S (1992) A new source of biologically active compounds–earthworm tissue (*Eisenia foetida*, *Lumbricus rubellus*). Comp Biochem Physiol 102A: 441–447
Jackson AD, Smith VJ, Peddie CM (1993) In vitro phenoloxidase activity in the blood of *Ciona intestinalis* and other ascidians. Dev Comp Immunol 17: 97–108
Janda V, Bohuslav P (1934) Sur l'explantation du tissu de la paroi intistinale et des amibocytes de *Lumbricus terrestris* L. et des cellules d'épithelium intestinal d'*Anodonta cygnea* L. Publ Fac Sci Univ Charles 133: 1–23 (in Czech with French Summary)
Janeway CA (1992) The immune system evolved to discriminate infectious nonself from noninfectious self. Immunol Today 13: 11–16
Jollès P, Zuili S (1960) Purification et étude comparée de nouveaux lysozymes extraits du poumon de poule et de *Nephthys hombergi*. Biochim Biophys Acta 39: 212–217
Karp RD (1985) Preliminary characterization of the inducible humoral factor in the American cockroach (*Periplaneta americana*). Dev Comp Immunol 9: 569–575
Kauschke E, Mohrig W (1987) Comparative analysis of hemolytic and hemagglutinating activities in the coelomic fluid of *Eisenia foetida* and *Lumbricus terrestris* (Annelida, Lumbricidae). Dev Comp Immunol 11: 331–342
Keilin ND (1925) Parasitic autotomy of the host as a mode of liberation of coelomic parasites from the body of the earthworm. Parasitology 17: 170–172
Komano H, Kasama E, Nagasawa Y, Nakanishi Y, Matsuyama K, Ando KI, Natori S (1987) Purification of *Sarcophaga* (fleshfly) lectin and detection of sarcotoxins in the culture medium of NIH-Sape-4, an embryonic cell line of *Sarcophaga peregrina*. Biochem J 248: 217–222
Lambert J, Keppi E, Dimarqo JL, Wicker C, Reichhart JM, Dunbar B, Lepage P, Van Dorsselaer A, Hoffmann J, Fothergill J, Hoffmann D (1989) Insect immunity: isolation from immune blood of the dipteran *Phormia terranovae* of two insect antibacterial peptides with sequence homology to rabbit lung macrophage bactericidal peptides. Proc Natl Acad Sci USA 86: 262–266
Langone JJ (1982) Protein A and related receptors. Adv Immunol 32: 158–241
Lassalle F, Lassègues M, Roch P (1988) Protein analysis of earthworms coelomic fluid–IV. Evidence, activity, induction and purification of *Eisenia fetida andrei* lysozyme (Annelidae). Comp Biochem Physiol 91B: 187–192
Laulan A, Morel A, Lestage J, Delaage M, Chateaureynaud-Duprat P (1985) Evidence of synthesis by *Lumbricus terrestris* of specific substance in response to an immunization with a synthetic hapten. Immunology 56: 751–758
Laulan A, Lestage J, Bouc AM, Chateaureynaud-Duprat P (1988) The phagocytic activity of *Lumbricus terrestris* coelomocytes is enhanced by the vertebrate opsonins: IgG and complement C3b fragment. Dev Comp Immunol 12: 269–278

Leipner C, Tučková L, Rejnek J, Langner J (1993) Serine proteases in coelomic fluids of annelids *Eisenia foetida* and *Lumbricus terrestris*. Comp Biochem Physiol 105B: 637–641

Marchalonis J, Atwell JL, Goding JW (1978) 7S immunoglobulins of monotreme, the Echidna *Tachyglossus acutaetus*: two distinct isotypes which bind A protein of *Staphylococcus aureus*. Immunology 34: 97–103

Metchnikoff EE (1887) Sur la lutte des cellules de l'organisme contre l'invasion des microbes. Ann Inst Pasteur 1: 322–340

Mihara H, Sumi H, Yoneta T, Mizumoto H, Ikeda R, Seiki M, Maruyama M (1991) A novel fibrinolytic enzyme extracted from the earthworm *Lumbricus rubellus*. Jpn J Physiol 41: 461–472

Mohrig W, Kauschke E, Ehlers M (1984) Rosette formation of the coelomocytes of the earthworm *Lumbricus terrestris* L. with sheep erythrocytes. Dev Comp Immunol 8: 471–476

Nagasawa H, Sawaki K, Fujii Y, Kobayashi M, Segawa T, Suzuki R, Inatomi H (1991) Inhibition by lombricine from earthworm (*Lumbricus terrestris*) of the growth of spontaneous mammary tumours in SHN mice. Anticancer Res 11: 1061–1064

Olive PJW (1974) Cellular aspects of regeneration influence in *Nereis diversicolor*. J Embryol Exp Morphol 32: 111–131

Périn JP, Jollès P (1972) The lysozyme from *Nephthys hombergi* (annelid). Biochim Biophys Acta 263: 683–689

Peucellier G (1983) Purification and characterization of proteases from the polychaete annelid *Sabellaria alveolata*. Eur J Biochem 136: 435–445

Porchet-Henneré E, M'Berri M (1987) Cellular reaction of the polychaete annelid *Nereis diversicolor* against coelomic parasites. J Invertebr Pathol 50: 58–66

Ratcliffe NA, Leonard C, Rowley AF (1984) Prophenoloxidase activation: nonself recognition and cell cooperation in insect immunity. Science 226: 557–559

Ratcliffe NA, Rowley AF, Fitzgerald SW, Rhodes CP (1985) Invertebrate immunity: basic concepts and recent advances. Int Rev Cytol 97: 183–349

Ratcliffe NA, Brookman JL, Rowley AF (1991) Activation of the prophenoloxidase in locusts by bacterial lipopolysaccharides. Dev Comp Immunol 15: 33–39

Rejnek J, Tučková L, Šíma P, Kostka J (1986) The proteins in *Lumbricus terrestris* and *Eisenia foetida* coelomic fluids and on coelomocytes reacting with sheep and goat IgG molecules. Dev Comp Immunol 10: 467–475

Rejnek J, Tučková L, Zikán J, Tomana M (1991) The interaction of a protein from the coelomic fluid of earthworms with staphylococcal protein A. Dev Comp Immunol 15: 269–277

Rejnek J, Tučková L, Šíma P, Bilej M (1993) The fate of protein antigen in earthworms: study in vivo. Immunol Lett 36: 131–136

Richman DD, Cleveland PH, Oxman MN, Johnson KM (1982) The binding of staphylococcal protein A by the sera of different species. J Immunol 128: 2300–2305

Roch P (1977) Rèactavité in vitro des leucocytes du lombricien *Eisenia fetida* Sav. a quelques substance mitogéniques. CR Acad Sci Ser D 284: 705–712

Roch P (1979) Protein analysis of earthworm coelomic fluid: I-polymorphic system of the natural hemolysin of *Eisenia fetida andrei*. Dev Comp Immunol 3: 599–608

Roch P, Cooper EL (1983) A β_2-microglobulin-like molecule on earthworm (*L. terrestris*) leukocyte membranes. Dev Comp Immunol 7: 633–636

Roch P, Valembois P, Du Pasquier L (1975) Response of earthworm leukocytes to concanavalin A and transplantation antigens. In: Hildemann WH, Benedict AA (eds) Immunologic phylogeny. Plenum Press, New York, pp 45–54

Roch P, Valembois P, Davant N, Lassègues M (1981) Protein analysis of earthworm coelomic fluid. II. Isolation and biochemical characterisation of the *Eisenia fetida andrei* factor (EFAF). Comp Biochem Physiol 69B: 829–836

Roch P, Cooper EL, Eskinazi DP (1983) Serological evidences for a membrane structure related to human β_2-microglobulin expressed by certain earthworm leukocytes. Eur J Immunol 13: 1037–1042

Roch P, Davant N, Lassègues M (1984) Isolation of agglutinins from lysins in earthworm coelomic fluid by gel filtration followed by chromatofocusing. J Chromatogr 290: 231–235

Roch P, Lassègues M, Valembois P (1991a) Antibacterial activity of *Eisenia fetida andrei* coelomic fluid. III. Relationship within the polymorphic hemolysins. Dev Comp Immunol. 15: 27–32
Roch P, Stabili L, Pagliara P (1991b) Purification of three serine proteases from the coelomic cells of earthworms (*Eisenia fetida*). Comp Biochem Physiol 98B: 597–602
Schrevel J (1969) Recherches sur le cycle des Lucodinidae Grégarines parasites d'annélides polychètes. Protistologica 5: 561–588
Schrevel J (1970) Contribution à l'étude des Selenidiidae parasites d' annélides polychètes. I. Cycles biologiques. Protistologica 6: 389–426
Schrevel J (1971) Contribution à l' étude des Selenidiidae parasites d' annélides polychètes. II. Ultrastructure de quelques trophozoites. Protistologica 7: 439–450
Schubert I, Messner B (1971) Untersuchungen über das Vorkommen von Lysozym bei Anneliden. Zool Jahrb Physiol 76: 36–50
Shalev A, Goldenberg PZ, Huebner E (1980) Evidence for an H-Y cross-reactive antigen in invertebrates. Differentiation 16: 77–80
Shalev A, Greenberg AH, Logdberg L, Bjorck L (1981) β_2-Microglobulin-like molecules in low vertebrates and invertebrates. J Immunol 127: 1186–1191
Shalev A, Segal S, Eli MB (1985) Evolutionary conservation of brain Thy-1 glycoprotein in vertebrates and invertebrates. Dev Comp Immunol 9: 497–506
Šinkora M, Bilej M, Tučková L, Romanovský A (1993) Hemolytic function of opsonizing protein of earthworm's coelomic fluid. Cell Biol Int 17: 935–939
Söderhäll K, Smith VJ (1986) The prophenoloxidase activating system: the biochemistry of its activation and role in arthropod cellular immunity, with special reference to crustaceans. In: Brehélin M (ed) Immunity in invertebrates. Springer, Berlin Heidelberg New York, pp 208–223
Stein E, Cooper EL (1981) The role of opsonins in phagocytosis by coelomocytes of the earthworm *Lumbricus terrestris*. Dev Comp Immunol 5: 415–425
Stein EA, Cooper EL (1983) Carbohydrate and glycoprotein inhibitors of naturally occurring and induced agglutinins in the earthworm *Lumbricus terrestris*. Comp Biochem Physiol 76B: 197–206
Stein EA, Cooper EL (1988) In vitro agglutinin production by earthworm leukocytes. Dev Comp Immunol 12: 531–548
Stein EA, Avtalion RR, Cooper EL (1977) The coelomocytes of the earthworm *Lumbricus terrestris*: morphology and phagocytic properties. J Morphol 153: 467–476
Stein EA, Wojdani A, Cooper EL (1982) Agglutinins in the earthworm *Lumbricus terrestris*: naturally occurring and induced. Dev Comp Immunol 6: 407–421
Stein EA, Younai S, Cooper EL (1986) Bacterial agglutinins of the earthworm, *Lumbricus terrestris*. Comp Biochem Physiol 84B: 409–415
Stein EA, Younai S, Cooper EL (1990) Separation and partial purification of agglutinins from coelomic fluid of the earthworm, *Lumbricus terrestris*. Comp Biochem Physiol 97B: 701–705
Sun SC, Lindström I, Boman HG, Faye I, Schmidt O (1990) Hemolin: an insect-immune protein belonging to the immunoglobulin superfamily. Science 250: 1729–1732
Takahashi T, Iwase T, Kobayashi K, Rejnek J, Mestecky J, Moro I (1992) Phylogeny of the immunoglobulin joining (J) chain. 7th Int Congr Mucosal Immunology, Prague, Czechoslovakia, 16–20 Aug 1992, Czechoslovak Immunological Society, Prague, 234 pp
Toupin J, Lamoureux G (1976) Coelomocytes of earthworms: phytohemagglutinin (PHA) responsiveness. In: Wright RK, Cooper EL (eds) Phylogeny of thymus and bone marrow-bursa cells. Elsevier, Amsterdam, pp 19–27
Tučková L, Bilej M (1994) Antigen processing in earthworms. Immunol Lett 41: 273–277
Tučková L, Rejnek J, Šíma P, Ondřejová R (1986a) Lytic activities in coelomic fluids of *Eisenia foetida* and *Lumbricus terrestris*. Dev Comp Immunol 10: 181–189
Tučková L, Rejnek J, Šíma P (1986b) Lytic activites in coelomic fluid of annelids *E. foetida* and *L. terrestris*. 6th Int Congr Immunol Toronto, National Research Council Canada, 1: 52.4 (Abstr)
Tučková L, Rejnek J, Šíma P (1988) Response to parenteral stimulation in earthworms *L. terrestris* and *E. foetida*. Dev Comp Immunol 12: 287–296

Tučková L, Rejnek J, Bilej M, Pospíšil R (1991a) Characterization of antigen-binding protein in earthworms *Lumbricus terrestris* and *Eisenia foetida*. Dev Comp Immunol 15: 263–268
Tučková L, Rejnek J, Bilej M, Hájková H, Romanovský A (1991b) Monoclonal antibodies to antigen binding protein of annelids (*Lumbricus terrestris*). Comp Biochem Physiol 100B: 19–23
Vaillier J, Cadoret MA, Roch P, Valembois P (1985) Protein analysis of earthworm coelomic fluid. III. Isolation and characterization of several bacteriostatic molecules from *Eisenia foetida andrei*. Dev Comp Immunol 9: 11–20
Valembois P (1963) Recherches sur la nature de la réaction antigreffe chez le lombricien *Eisenia foetida*. C R Acad Sci D (Paris) 257: 3489–3490
Valembois P, Roch P, Du Pasquier L (1973) Dégradation in vitro de protéine éntrangére par les macrophages du Lombricien *Eisenia fetida* Sav. C R Acad Sci Paris Sér III 277: 57–60
Valembois P, Roch P, Boiledieu D (1980) Natural and induced cytotoxicities in sipunculids and annelids. In: Manning MJ (ed) Phylogeny of immunological memory. Elsevier, Amsterdam, pp 47–55
Valembois P, Roch P, Lassègues M, Cassand P (1982) Antibacterial activity of the hemolytic system from the earthworm *Eisenia fetida andrei*. J Invertebr Pathol 40: 21–27
Valembois P, Roch P, Lassègues M (1986) Antibacterial molecules in annelids. In: Brehélin M (ed) Immunity in invertebrates. Springer, Berlin Heidelberg New York, pp 74–93
Valembois P, Seymour J, Roch P (1991) Evidence and cellular localization of an oxidative activity in the coelomic fluid of the earthworm *Eisenia fetida andrei*. J Invertebr Pathol 57: 177–183
Valembois P, Lassègues M, Roch P (1992) Formation of brown bodies in the coelomic cavity of the earthworm *Eisenia fetida andrei* and attendant changes in shape and adhesive capacity of constitutive cells. Dev Comp Immunol 16: 95–101
Větvička V, Šíma P, Cooper EL, Bilej M, Roch P (1994) Immunology of annelids. CRC Press, Boca Raton
Villaro AC, Sesma P, Alegría D, Vázquez JJ, López J (1985) Relationship of symbiotic microorganisms to metanephridium: phagocytic activity in the metanephridinal epithelium of two species of Oligochaeta. J Morphol 186: 307–314
Vivier E, Henneré E (1964) Cytologie, cycle et affinités de la Coccidie Coelotropha durchoni nomen novum (= *Eucoccidium durchoni* Vivier), parasite de *Nereis diversicolor* O. F. Müller (Annélide Polychète). Bull Biol Fr Belg 1: 154–206
Voburka Z, Mareš M, Větvička V, Bilej M, Baudyš M, Fusek M (1992) New trypsin inhibitors are present in the coelomic fluid of the earthworm *Lumbricus terrestris*. Biochem Int 27: 679–685
Wojdani A, Stein EA, Lemmi CA, Cooper EL (1982) Agglutinins and proteins in the earthworm, *Lumbricus terrestris*, before and after injection of erythrocytes, carbohydrates, and other materials. Dev Comp Immunol 6: 613–624
Zikán J, Šíma P, Prokešova L, Hadge D (1980) Binding of nonmammalian immunoglobulins to staphylococcal protein A. Folia Biol (Prague) 26: 261–266
Zwilling R, Neurath H (1981) Invertebrate proteases. In: Lorand L (ed) Methods in enzymology, vol 80. Academic Press, New York, pp 633–664

Cell Products: Natural and Induced as Revealed by Non-specific and Specific Responses Following Antigenic Challenge

Chapter 3

The Prophenoloxidase Activating System: A Common Defence Pathway for Deuterostomes and Protostomes?

V.J. Smith

Contents

1 Introduction

For many years, invertebrate immunology has had at its core fundamental questions about the nature of non-self recognition, the mechanisms of cell activation and the existence of common defence pathways and/or factors. These questions are rooted in our desire to trace the phylogeny of immunity and to find the evolutionary origins of the vertebrate immune system. Modern reductionist

School of Biological and Medical Sciences, Gatty Marine Laboratory, University of St Andrews, Fife, Scotland KY16 8LB, UK

Advances in Comparative and Environmental Physiology, Vol. 23

thinking, which perceives the animal kingdom as a continuum of structural, physiological and biochemical development, holds that complex processes or molecules must have precursors in more ancient groups and that certain (presumably 'primitive') factors or pathways will be present across a range of phyla. An understanding of the defence molecules or responses in invertebrates should therefore shed some light on the primordial state of the mammalian immune network.

Traditionally, comparative assessments of immune capability have taken a top-down approach, i.e. have attempted to detect vertebrate-type factors or responses in extant invertebrate species. In many respects this is a logical approach, but it has the drawback of assuming that molecules of functional importance in vertebrates will also be important in invertebrates. It also assumes that the activities of key factors in vertebrates will be expressed in a similar way in invertebrates. Put simply, the top-down view is inherently vulnerable to 'vertebrate bias'. Early work, for example, concentrated on searching for antibody-type factors or activities in invertebrates, and there was much effort in the 1960s and 1970s to identify adaptive type reactivity in invertebrate host defence. It is now widely accepted that invertebrates do not produce immunoglobulin-type molecules or classical complement factors and are unlikely to show adaptive immune responses based on a clonally derived recognition system (Klein 1989).

An alternative approach is to compare immune capability in a bottom-up manner, i.e. to seek defence molecules or activities in a range of protostome phyla and to look for similar factors in deuterostome animals. It is not implausible that vertebrates might have retained the vestiges of primordial defence systems from their invertebrate ancestors (Janeway 1989), although the discovery of conserved proteins or processes is likely to be due more to serendipity than heuristic endeavour. Accordingly few investigations have adopted this approach, but where it has been applied, it has often yielded valuable information. A good example is the work of Boman and his colleagues on antibacterial peptides in insects (see Boman 1991). Certainly, deuterostome invertebrates rely on inflammatory cellular reactions, such as phagocytosis, encapsulation and the production of antimicrobial factors for host defence in much the same way as protostome animals (see reviews by Ratcliffe et al. 1985; Smith 1991). Therefore, there may be an element of equivalence in the biochemical agents they use to induce and regulate cellular activity. Factors common to both groups are likely to show some degree of biochemical and/or molecular homology, irrespective of their biological roles. We are gradually coming to understand the biochemical basis underlying recognition and blood cell activation in invertebrates, and with the tools of molecular biology at our disposal, evaluation of similar pathways or factors in different phyla is now possible.

This chapter aims to take a bottom-up approach in assessing the extent to which one defence system in protostome animals might be distributed across the various invertebrate groups. This pathway, the so-called prophenoloxidase activating (proPO) system, was proposed, some ten years ago, to represent a defence and/or recognition system for arthropods (Söderhäll 1982; Söderhäll and Smith 1986). It attracted attention because it appeared to have certain biochemical and functional similarities to the alternate pathway of complement (Söderhäll and Smith 1986). There can be little doubt that the proPO system is closely linked to host defence in insects and crustaceans, but its status as an 'invertebrate complement' is probably less important than an appreciation of how the system arose and whether key factors have been conserved through evolution. Certainly, melanin, the final product of phenoloxidase activity, is ubiquitous throughout the animal kingdom and occurs in nearly all groups from sponges to mammals. The pigment is also present in some fungi, bacteria and plants, so it is not improbable that the biochemical machinery responsible for the activation of phenoloxidase might similarly have a wide phylogenetic distribution. The present review therefore assesses the presence of proPO-like components in non-arthropod invertebrates, examines the extent of homology between factors found in protostomes and deuterostomes and considers the role that any proPO components might have in host defence and/or non-self recognition in non-arthropod species.

2 The Prophenoloxidase Activating System in Arthropods

That melanization accompanies haemocytic reactivity to foreign agents in arthropods has been known for many years, but clues to its link with cell activation only emerged with the discovery that bacteria or soluble carbohydrates from fungal cell walls induce melanin formation by insect or crustacean blood in vitro (Pye 1974; Unestam and Söderhäll 1977). The key factor in this reaction is the enzyme phenoloxidase, which usually exists in arthropods in the inactive form of prophenoloxidase. Conversion of prophenoloxidase to its active state is achieved by proteolytic cleavage and the resulting enzyme catalyses both the *o*-hydroxylation of monophenols and the oxidation of diphenols to quinones. In turn, these quinones are polymerized non-enzymically to melanin. Activation of prophenoloxidase itself is dependent upon a cascade of serine proteases and other factors in the blood, and some of these factors are sensitive to lipopolysaccharides (LPS), β-1,3-glucans or other microbially derived carbohydrates. Upon activation, the component proteins participate in a range of biological responses, including phagocytosis, encapsulation, clotting, microbial killing and wound repair. The biochemical features and functional effects of prophenoloxidase activation in arthropods have been comprehensively reviewed by Söderhäll (1982), Söderhäll and Smith (1986), Smith and Söderhäll (1986), Johansson and Söderhäll (1989a), Söderhäll et al. (1990) and Söderhäll and Cerenius (1992), so the details will not be reiterated here. Since it is generally accepted that the proPO system finds its full expression in the arthropods, the present account will, instead, concentrate on describing key factors associated with the activation sequence in order to compare them with equivalent molecules in other invertebrate phyla.

2.1 Crustaceans

2.1.1 Component Proteins of the Activation Sequence and Related Factors

The most extensive studies of prophenoloxidase activation have been made in freshwater crayfish, chiefly *Astacus astacus* and *Pacifastacus leniusculus* (Söderhäll and Smith 1986; Smith and Söderhäll 1986, Johansson and Söderhäll 1989a; Söderhäll et al. 1990; Table 1). However, phenoloxidase activity has been reported to occur in the blood of a wide range of marine crustaceans (Smith and Söderhäll 1983a, 1991; Söderhäll and Smith 1983; Rolle et al. 1990) and in nearly every case, it has been found to reside in the haemocytes, particularly the granular and semi-granular cells. Typically, phenoloxidase activity is measured in vitro by the oxidation of L-dopa at neutral pH, and for most species activity is strongly enhanced by exogenous proteases or bacterial LPS (Smith and Söderhäll 1983a, b, 1991; Söderhäll and Smith 1983). While the degree of enhancement varies from species to species (Smith and Söderhäll 1991), general features of enzyme activity are dependence upon divalent cations, sensitivity to serine protease inhibitors, e.g. soybean trypsin inhibitor (STI) or benzamidine, and impairment by phenyl thiourea (PTU) (Smith and Söderhäll 1983a; Ashida and Söderhäll 1984; Söderhäll et al. 1986).

With crayfish, a number of proteins associated with proPO activation, including prophenoloxidase itself, have now been purified and characterized (Table 1), and a clearer picture of the ways these factors interact is beginning to emerge. Amongst the various factors of the proPO system, prophenoloxidase has come to be regarded as a key 'marker' protein, although its purification has been complicated by its sensitivity to non-physiological activation by solvents, detergents and other agents or treatments. By using melittin from bee venom to stabilize the enzyme from crayfish blood, Aspán and Söderhäll (1991) have obtained an inactive preparation, homogeneous on denaturing polyacrylamide gels. The purified factor is a glycoprotein with a molecular mass of 76 kDa, an isoelectric point of ca. 5.4 and a copper content of ca. 0.16% (Aspán and Söderhäll 1991). The enzyme is cleaved by trypsin to yield a single phenoloxidase with a molecular mass of 60 kDa (Aspán and Söderhäll 1991). By contrast, treatment with an endogenous blood cell protease results in the production of two forms of active phenoloxidase with molecular masses of 62 and 60 kDa, respectively (Aspán and Söderhäll 1991). Blood cell phenoloxidase from crayfish seems to differ from cuticular phenoloxidases described for other crustacean species, where the molecular mass has been estimated to range from 82–97 kDa for spiny lobsters (Chen et al. 1991) and 30–40 kDa for shrimps (Simpson et al. 1988; Table 1).

Another factor associated with proPO activation, purified from crayfish blood cells, is a serine protease, designated ppA (Aspán et al. 1990). This has a molecular mass of ca. 36 kDa, exhibits maximum activity at neutral or slightly alkaline pH and is inactivated by heating to 58 °C (Aspán et al. 1990). Three other serine proteases, with molecular masses of 38, 50 and 67 kDa, respectively, co-exist with ppA in crayfish blood cells (Aspán and Söderhäll 1991), but only the 38 and 36 kDa enzymes cleave the chromogenic peptide substrate S-2337 (Bz-Ile-Glu-(γ-*o*-Piperidyl)-Gyl-Arg-*p*-nitroaniline) and specifically bind to prophenoloxidase in vitro (Aspán and Söderhäll 1991). While both of these proteases are activated by β-1,3-glucans, purified ppA alone cleaves a homogeneous prophenoloxidase preparation without additional factors (Aspán and Söderhäll 1991). Consequently, ppA has been proposed as the native protease responsible for activating prophenoloxidase in vivo (Aspán and Söderhäll 1991).

As far as the release and subsequent activation of proPO factors from the blood cells is concerned, work on crayfish has indicated that a plasma factor might be involved in the response. This factor binds to β-1,3-glucans and is termed the glucan-binding protein (βGBP) (Duvic and Söderhäll 1990). It is a monomeric glycoprotein with a molecular mass of ca. 100 kDa (Duvic and Söderhäll 1990). The factor does not appear to be an agglutinin and has no proteolytic or oxidative activity of its own (Duvic and Söderhäll 1990). However, in the presence of (specifically) β-1,3-glucans, it enhances both haemocyte-derived protease and phenoloxidase activities in vitro (Duvic and Söderhäll 1990). An antigenically identical molecule also seems to be present in crayfish blood cells (Duvic and Söderhäll 1990) indicating that the protein might be synthesized by the haemocytes before release into the plasma.

Regarding exocytosis of proPO factors from the blood cells, Smith and Söderhäll (1983b) have shown that crayfish semigranular cells undergo degranulation in vitro upon exposure to bacteria. Equivalent exocytosis of the granular cells is achieved only by an active proPO system (Smith and Söderhäll 1983b). This response of the granular cells seems to be due, at least in part, to a 76 kDa protein produced during proPO activation (Johansson and Söderhäll 1989b). Since this factor appears to be synthesized in the haemocytes and stored in the cytoplasmic granules (Liang et al. 1992), other mechanisms must operate to bring about its release from the cells at the appropriate time. A possible candidate is the 100 kDa glucan-binding protein (βGBP) from crayfish plasma (see above), which after complexation with laminarin (a β-1,3-glucan), has been found to bind to and stimulate exocytosis of the granular cells in vitro (Barracco et al. 1991). Neither laminarin nor βGBP alone have a similar exocytotic effect and do not bind to the haemocyte surface (Barracco et al. 1991). Thus, it appears that release of the proPO system from the cells is a regulated process entailing at least two factors, one in plasma, the other within the haemocytes.

Amongst the various factors which are associated with the activation of prophenoloxidase in crayfish blood are a number of protease inhibitors that may serve either to regulate the activating cascade or to help protect the host against invasive parasites (Häll and Söderhäll 1982; Hergenhahn et al. 1987, 1988; Hall et al. 1989; Aspán et al. 1990). One of these is a 190 kDa dimer with α_2-macroglobulin-like properties (Hergenhahn et al. 1988). It is believed to cage proteases in a similar way to α_2-macroglobulin of vertebrates, and may therefore control amplification of proPO activation in the haemocoel, or, alternatively, neutralize aggressive proteases from parasites. Other inhibitors detected in crayfish blood are a 23 kDa subtilisin inhibitor (Häll and Söderhäll 1982) and a 155 kDa trypsin inhibitor (Hergenhahn et al. 1987).

Table 1. Summary of constituent and associated factors of the proPO system in Arthropods

Factor	Function	Molecular Mass (kDa)	Location	Species	References
Prophenoloxidase (proPO)	Stable enzyme which upon activation cleaves phenols to quinones during melanization	76	Haemocytes	*Pacifastacus leniusculus*	Aspán and Söderhäll (1991)
		80	Plasma	*Bombyx mori*	Ashida (1971)
		76	Haemolymph	*Hyalophora cecropia*	Andersson et al. (1989)
		100–150	Haemolymph	*Musca domestica*	Aso et al. (1985)
		178	Haemolymph	*Calliphora vicinia*	Tsukamoto et al. (1986)
		87	Haemolymph	*Calliphora erythrocephala*	Naqvi and Karlson (1979)
		115	Haemolymph		Munn and Bufton (1973)
Phenoloxidase (PO)	Active form of proPO	82–97	Cuticle	Spiny lobsters	Chen et al. (1991)
		30–40	Cuticle	Shrimps	Simpson et al. (1988)
		60	Haemocytes	*P. leniusculus*	Aspán and Söderhäll (1991)
		70	Plasma	*B. mori*	Ashida and Yoshida (1988)
		70	Cuticle	*B. mori*	Ashida et al. (1983)
		310–340	Cuticle	*M. domestica*	Tsukamoto et al. (1986)
		220	Haemolymph	Millipedes	Xylander (1992a)
Activating enzyme (ppA)	Activates proPO	35	Haemocytes	*P. leniusculus*	Aspán et al. (1990)
		67 (dimer)	Haemolymph	*B. mori*	Dohke (1973), Ashida and Dohke (1980)
					Andersson et al. (1989)
Other proteases	Bring about activation of ppA	33, 35	Haemolymph	*H. cecropia*	Munn and Bufton (1973)
		115	Cuticle	*Calliphora* spp.	
Glucan-binding protein (βGBP)	Binds to β-1, 3-glucans may act as an opsonin.	100	Plasma	*P. leniusculus*	Duvic and Söderhäll (1990)
		90	Plasma	*Blaberus craniifer*	Söderhäll et al. (1988)
		62	Plasma	*B. mori*	Ochiai and Ashida (1988)
					Thörnqvist et al. (1994)

LPS-binding protein	Binds to lipopolysaccharides	28	Haemolymph	*Periplaneta americana*	Jomori et al. (1990)
		123	Amoebocytes	*Limulus polyphemus*	Nakamura et al. (1986)
		15	Amoebocytes	*L. polyphemus*	Morita et al. (1985)
Cell adhesion factor (CAF)	Causes granular cells to exocytose and adhere to surfaces, possible opsonic	76	Haemocytes	*P. leniusculus*	Johansson and Söderhäll (1988)
		90	Haemocytes	*B. craniifer*	Rantamäki et al. (1991)
		55	Amoebocytes	*L. polyphemus*	Liu et al. (1991)
Protease inhibitors (incl. $\alpha_2$macroglobulin)	Block the activity of proteases, may regulate proPO activation	190	Plasma	*P. leniusculus*	Hergenhahn et al. (1988)
		112	Haemolymph	*H. cecropia*	Andersson et al. (1989)
		12	whole animal extract	*Drosophila melanogaster*	Kang and Fuchs (1980)
	Inhibits subtilisin	23	Cuticle	*P. leniusculus*	Hall and Söderhäll (1982)
	inhibits trypsin	155	Plasma	*P. leniusculus*	Armstrong et al. (1990)
	α_2m	185	Amoebocytes	*L. polyphemus*	
Clotting proteins	Forms stable clot in cell lysate extracts	420 (dimer)	Plasma	Spiny lobsters	Fuller and Doolittle (1971a)
		210	Plasma	*P. leniusculus*	Kopacek et al. (1993)
		84	Amoebocytes	*L. polyphemus*	Sullivan and Watson (1975)

The latter inhibits ppA and is thought to modulate proPO activation in the haemocoel (Aspan et al. 1990). It is unclear whether equivalent factors occur in related crustacean species, although a thioester-containing α_2-macroglobulin homologue has been isolated from lobster, *Homarus americanus*, haemolymph (Spycher et al. 1987). As yet definitive evidence that this regulates proPO activation for lobsters has not been provided.

2.1.2 Biological Functions

A variety of defensive roles have been attributed to the proPO system in crustaceans (Table 2). Evidence for this is somewhat circumstantial and has been based largely on the observed effects of proPO elicitors, or haemocyte lysate supernatants containing active proPO factors, or on the host responses to foreign materials in vitro or in vivo (Smith and Söderhäll 1983a; Smith et al. 1984; Söderhäll et al. 1986). Cellular strategies which have been linked to prophenoloxidase activation in crustaceans in this way include the generation of opsonins (Smith and Söderhäll 1983a; Söderhäll et al.1986), stimulation of encapsulation (Smith et al. 1984; Söderhäll et al. 1984; Persson et al. 1987), clotting (Söderhäll 1981) and release of antimicrobial factors (Söderhäll and Ajaxon 1982). One factor, termed the cell adhesion factor (CAF), has

Table 2. Host defence activities or responses and the proPO system in different arthropod groups

Activity	Crustaceans	Insects	Chelicerates	Myriapods
Phenoloxidase activity (PO)	Yes	Yes	No	Yes
Protease activity	Yes, related to PO activity	Yes, related to PO activity	Yes, linked to clotting	NK
Protease inhibitors	Yes, several including α_2 macroglobulin	Yes	Yes, including α_2 macroglobulin	NK
Degranulation	Yes	Yes	Yes, linked to clotting	Yes, but link to proPO not confirmed
Cell adhesion	Yes	Yes	Yes, probably associated with clotting	Yes, but link to proPO not confirmed
Opsonization	Yes	Yes, but factor(s) not identified	NK	NK
Microbial killing	Yes, link to proPO not confirmed	Yes, link to proPO unknown	Yes, probably associated with clotting	Yes, but link to proPO not confirmed
Clotting	Yes, induced by microbial carbohydrates, linked to proPO	Yes, probably induced by microbial carbohydrates	Yes, induced by microbial carbohydrates	NK

For references, see text. NK, not known.

been purified from *P. leniusculus* and found to mediate the attachment and spreading of the haemocytes in vitro (Johansson and Söderhäll 1988). As mentioned above, this 76 kDa protein is produced simultaneously with activation of the proPO system and causes degranulation when added to monolayers of the granular cells (Johansson and Söderhäll 1989b). In many animals, cell adhesion is known to be influenced by particular basement membrane proteins. One of them, fibronectin, contains the peptide triplet RGD. Since the biological effect of CAF for crayfish haemocytes is mimicked by the peptide Gly-Arg-Gly-Asp-Ser (GRGDS), but not other peptides with different amino acid sequences, it is possible that, in common with vertebrates, caryfish cells possess receptors capable of recognizing the RGD sequence (Johansson and Söderhäll 1989c). Moreover, there is some evidence that the glucan-binding protein (*β*GBP) in crayfish plasma might further contribute to the response since it stimulates spreading of the granular cells after reaction with *β*-1,3-glucans (Barracco et al. 1991).

The 76 kDa CAF has also been suggested to possess characteristics which would enable it to function as an opsonin (Johansson and Söderhäll 1989a). Specifically, CAF may promote the adhesion of bacteria or other non-self particles to the phagocyte surface and hence facilitate their subsequent ingestion by the cell. Recently, evidence that the 76 kDa protein enhances uptake of yeast by the phagocytes of *Carcinus maenas* has been produced with similar opsonic activity shown for *β*GBP and a 80 kDa protein from the granular cells (Thörnqvist et al. 1994). Certainly, in crab, *Carcinus maenas*, phenoloxidase itself does not appear to have opsonic properties (Söderhäll et al. 1986), but because serine protease inhibitors which block LPS or *β*-1,3-glucan activation of prophenoloxidase also abolish the stimulatory effect of haemocyte lysate supernatants for crab phagocytes in vitro (Fig. 1; Bell and Smith, unpubl.), participation of other components of the proPO activating system cannot be ruled out.

Regarding clotting, it is well established that in crustaceans, the formation of a stable clot requires the interaction of the plasma with substances released from the haemocytes (Durliat 1985). The plasma clotting factor is probably functionally equivalent to fibrinogen of vertebrates, and a clotting protein has been purified from lobster, *Panulirus interruptus* by Fuller and Doolittle (1971a). Partial characterization has shown that it is a 420-kDa lipoglycoprotein composed of two identical disulphide-bonded subunits (Fuller and Doolittle 1971a). Polymerization is mediated by calcium-dependent transglutaminase activity in the blood cells (Fuller and Doolittle 1971b), and Kopacek et al. (1993) have isolated and characterized a clotting protein from crayfish, *P. leniusculus*, plasma. This similarly comprises two (210 kDa) subunits and forms stable clots in the presence of haemocyte lysate supernatants containing endogenous calcium dependent transglutaminase activity (Kopacek et al. 1993). The N-terminal amino acid sequence of the crayfish protein resembles that of *P. interruptus* indicating that the two molecules are homologous (Kopacek et al. 1993). In crayfish, the haemocytes contain clotting factors which are sensitive to microbial carbohydrates and gelation of haemocyte lysate supernatants involves one or more serine proteases (Söderhäll 1981). Thus, there are good grounds to believe that clotting is closely linked to proPO activation in these animals.

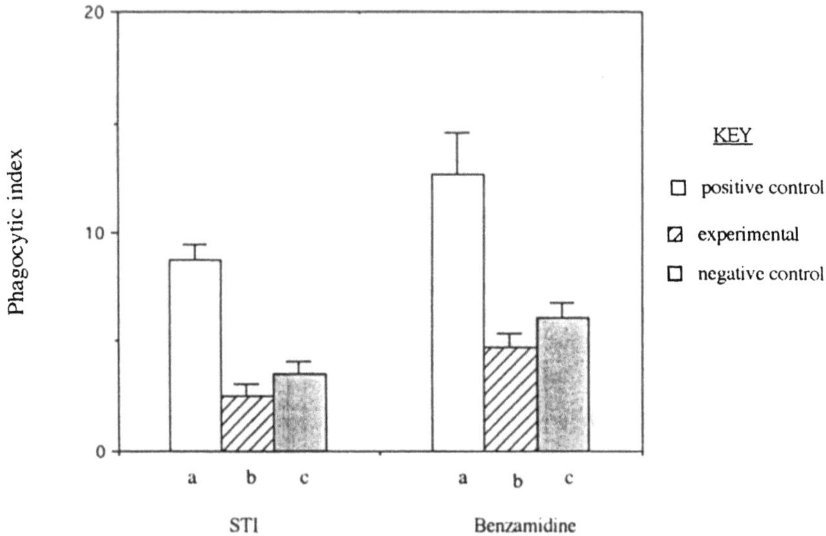

Fig. 1. Effect of serine protease inhibitors on the in vitro phagocytosis of opsonized bacteria by haemocytes of the shore crab, *Carcinus maenas*. (K.L. Bell, unpubl.). Three groups of monolayers of separated phagocytes from *C. maenas* were overlaid with washed *Psychrobacter immobilis* ($1 \times 10^8\, ml^{-1}$) pretreated in different ways. The first group, **a** the positive controls received bacteria which had been pre-incubated in haemocyte lysate supernatant (HLS) containing an active proPO system and then washed before presentation to the phagocytes. The HLS was activated by incubation with 0.1% laminarin (15 min, 20 °C). The second group, **b** the experimentals received bacteria pre-incubated in HLS containing an inhibited proPO system. To achieve this the HLS was treated with STI (5 mM) or benzamidine (100 mM) for 15 min before reaction with laminarin (15 min). The third group, **c** the negative controls were given bacteria pre-incubated in HLS containing a stable proPO system, ie treated only with with marine saline for 30 min. All monolayers were then incubated for 1 h at 20 °C, washed, fixed and examined under phase contrast optics to determine the percentage of haemocytes containing one or more intracellular bacteria. The values given are means $\pm$ SEM; $n = 5$

Finally, with respect to antimicrobial activity, while melanin and its quinone precursors are known to have cytotoxic and microbicidal properties (Pawelek and Lerner 1978; Söderhäll and Ajaxon 1982; Nappi and Vass 1993) which may help protect against parasitic or pathogenic invasion, a clear link between proPO activation and the production of bactericidal factors in crustaceans has not been shown. With the crab, *C. maenas*, Chisholm and Smith (1992) have reported that the proPO-bearing granular cells possess strong antibacterial activity against a range of Gram-negative and Gram-positive bacteria in vitro. Using the Gram-negative marine bacterium, *Psychrobacter immobilis* as the main test particle, activity was further found to be 90% effective within 60 min, independent of divalent cations and heat-stable (Chisholm and Smith 1992). Antibacterial activity also occurs in several other crustacean species, including *Galathea strigosa*, *Nephrops norvegicus*, *Crangon crangon* and the Antarctic isopod, *Glyptonotus antarcticus* (Chisholm 1993), although none appears to be detectable in freshwater crayfish (Söderhäll, pers. comm.). While antibacterial activity in *C. maenas* varies according to the size of the haemocyte population (Chisholm and Smith 1994), it does not correlate with phenoloxidase activity within the haemocytes. It is therefore unlikely that this enzyme is directly responsible for the observed phenomenon (Chisholm and Smith 1992). Other work has indicated that the response is also not due to agglutination, but as some activity may be lost by adsorption to LPS, the antibacterial factor(s) may act by binding to the bacterial cell wall (Chisholm 1993). Preliminary purification of the active agents by gel filtration indicates that there are at least three antibacterial proteins in *C. maenas*; two have molecular weights of ca. 70 and 38 kDa respectively and are bacteriostatic in action, while the third appears to be a peptide of ca. 6.5 kDa. All are active against both Gram-positive and Gram-negative bacteria (Chisholm 1993; Schnapp and Smith, unpubl.). The extent to which these factors are related to proPO activation is unknown and work is currently underway to further characterize these factors and to determine their precise mode of action in vivo.

2.2 Insects

Many workers have demonstrated phenoloxidase activity within the blood of insects and a sizeable body of literature on prophenoloxidase activation in this group of invertebrates is starting to accumulate. It would be impossible to review all the work done on prophenoloxidase activation in insects here, so attention will focus on highlighting the extent to which the insect proPO system resembles, or differs from, that of crustaceans.

2.2.1 Factors Involved in Activation

Prophenoloxidase has been purified from a number of insect species, and values for molecular masses have been found to range from 76 to 178 kDa (Table 1). As with crustacean prophenoloxidase, the enzyme in insects is activated by proteases, bacteria or β-1,3-glucans (Ashida 1971; Ashida et al. 1983), and has a pH optimum of ca. 7–8 (Dunphy 1991). Active phenoloxidase in the silkworm, *Bombyx mori*, has been found to have a molecular mass of 70 kDa; a value not dissimilar to that determined for crayfish (Aspán and Söderhäll 1991). Enzyme activation in insects further resembles that in crustaceans in that it may be blocked by serine protease inhibitors or PTU (Ashida et al. 1983; Leonard et al. 1985a, b; Dunphy 1991). It also tends to be calcium-dependent, and in *B. mori* high (i.e. > 100 mM) calcium concentrations inhibit activity probably by stabilization of the enzyme against proteolytic cleavage (Ashida et al. 1983). A major difference between insect and crustacean haemolymph prophenoloxidase is that in all crustaceans studied so far, the enzyme is present within the haemocytes, usually the granular cells, whereas in insects it may reside in the haemocytes (Leonard et al. 1985a, b) or the plasma (Iwama and Ashida 1986; Saul and Sugumaran 1988) (Table 1). There is also some uncertainty about the sensitivity of insect proPO to LPS; some workers have reported a failure to activate phenoloxidase in insects (Pye 1974; Ashida 1981; Ashida et al. 1983; Ratcliffe et al. 1984) while other workers have shown a positive response to this carbohydrate (Saul and Sugumaran 1987, 1988; Brookman et al. 1989).

Less is known about the native activators of prophenoloxidase in insect haemolymph. Several workers, notably Leonard et al. (1985a) and Dularay and Lackie (1985) have obtained indirect information about protease activity within insect haemocytes and established that it has certain similarities to haemocyte protease activity in crayfish. It may thus serve to cleave prophenoloxidase in vivo in a manner akin to crustaceans. Little information is available about the biochemical identity of the protease(s) responsible, although Andersson et al. (1989) have identified two proteases in the cecropid moth, *Hyalophora cecropia*. These have molecular masses of 33 and 53 kDa, respectively, and seem to be involved in activation of prophenoloxidase. Proteases are also known to be present in insect cuticle (Munn and Bufton 1973; Naqvi and Karlson 1979; Aso et al. 1985) (Table 1), and one in *B. mori* has been purified and shown to be a dimer of 67 kDa (Dohke 1973). While the purified form of this enzyme has been found to activate pure prophenoloxidase from *B. mori* hemolymph (Ashida and Dohke 1980), it is unlikely that this is the only natural activator of the enzyme in insects.

There is a limited amount of information available about the mechanism by which bacteria, LPS or β-1,3-glucans bring about activation of the proPO system in insects. One study by Jomori et al. (1990) has reported the isolation of a LPS-binding factor in the haemolymph of the cockroach, *Periplaneta americana*. The protein responsible is composed of 10–30 subunits, each with a molecular mass of 28 kDa (Jomori et al. 1990), and while this molecule may participate in proPO activation, no link has yet been demonstrated. Two other studies have described β-1,3-glucan binding proteins in insect plasma. One, by Söderhäll et al. (1988), has shown that there is a 90-kDa glycoprotein in the plasma of the cockroach, *Blaberus craniifer*, which appears to promote activation of haemocyte protease and prophenoloxidase by β-1,3-glucans in vitro. The other, by Ochiai and Ashida (1988), has identified a 62 kDa 'recognition protein' in *B. mori*. For neither species has the amino acid composition been determined. In terms of their biological effect, these two proteins would seem to be analogous to the βGBP in crayfish (see above), but whether the glucan-binding proteins in insects elicit degranulation reactions in insect cells in a similar way to the βGBP in crayfish is unclear.

Lastly, regarding regulation of prophenoloxidase activation in insects, serine protease inhibitors have been reported to be present in the plasma of several species, notably, *B. mori* (Sasaki 1984; Eguchi and Shomoto 1985), the tobacco hornworm, *Manduca sexta* (Saul and Sugumaran 1986), the flesh fly, *Sarcophaga peregrina* (Suzuki and Natori 1985), the fruit fly, *Drosophila melanogaster* (Kang and Fuchs 1980) and *H. cecropia* (Andersson et al. 1989; Table 1). For *M. sexta* and *H. cecropia*, these factors seem to help control protease and/or phenoloxidase activity in the haemolymph, but the extent to which they form part of the proPO system in the other species has not been ascertained.

2.2.2 Functions and Physiological Roles

The deposition of melanin at sites of infective foci, during clotting and in association with haemocyte nodules and capsules in insects has been known since the 1960s. In recent years, this has been taken as indicative of a role for proPO activating factors in the cellular defences (see review by Ratcliffe et al. 1985; Söderhäll and Smith 1986). However, there are surprisingly few reports which demonstrate a clear relationship between factors associated with proPO activation and cellular responsiveness in these animals (Table 2). In general, only phagocytosis and nodule formation/encapsulation reactions have received detailed study and been shown to be influenced by proPO factors.

Ratcliffe et al. (1984) and Leonard et al. (1895b) have examined the role of proPO in phagocytosis and demonstrated that activation of prophenoloxidase by microbial carbohydrates generates opsonins for the phagocytes. Importantly, the phenomenon entails co-operation between the proPO-bearing granular cells and the phagocytes (Anggraeni and Ratcliffe 1991) in a similar way to that reported for crustaceans (Söderhäll et al. 1986). So far, the factor(s) involved have not been positively identified and the precise way that prophenoloxidase generates opsonins in insects is unknown.

As regards the participation of proPO proteins in encapsulation by insect haemocytes, Brookman et al. (1989) have investigated the ability of various bacterial cell wall components to induce nodule formation in locusts. These authors also undertook correlative analyses of the efficiency of these carbohydrates to promote phenoloxidase activity in vitro. Their results show that nodule formation is initiated by the cell walls of both Gram-negative and Gram-positive bacteria, and that peptidoglycan, especially the lysozyme-soluble fraction, and capsular polysaccharide are the most potent elicitors. Although the responses were found to differ between locust species, a positive link between nodule formation and proPO activation does appear to exist in these animals (Brookman et al. 1989). Unfortunately, the mechanism by which proPO factors direct cell behaviour needs clarification as Dularay and Lackie (1985) found that locust haemocytes fail to encapsulate Sepharose beads, even after coating with at least five proteins from a haemocyte lysate supernatant containing an active proPO system. Recently, Rantamäki et al. (1991) have identified a 90-kDa protein in cockroach, *B. craniifer* haemocytes which has similar antigenic properties to the 76-kDa CAF of crayfish (see above). The *B. craniifer* protein is believed to serve in cell co-operation since it enhances adhesion of cockroach haemocytes to substrata in vitro and induces cell degranulation by regulated exocytosis (Rantamäki et al. 1991). While this offers strong support that insects and crustaceans share common cell signalling molecules, similar proteins have yet to be identified in other insect species and much still needs to be resolved about cell co-operation and proPO activation during capsule formation in this group of arthropods.

With respect to antibacterial activity, insects are known to produce a range of antibacterial peptides, including the attacins, cecropins and defensins, upon wounding or bacterial stimulation (Boman and Hultmark 1987; Jarosz 1993).

The same treatments are known to initiate clotting and activation of proPO proteins (Söderhäll and Smith 1986; Söderhäll et al. 1990), but, as yet, a clear biochemical connection between antibacterial peptides and the proPO system has not been demonstrated. In other respects, the information available to date indicates that there is good reason to suppose that insects possess several proPO proteins in common with crustaceans and that the system has similar functional properties in the two groups (Table 2).

2.3 Chelicerates

Chelicerates, particularly the horseshoe crabs, have been well studied with respect to their blood biochemistry, and information about the presence or absence of proPO-like factors comes largely from work on their clotting pathway. Atypically for arthropods, melanization is not a process which is distinctly associated with wounding, cellular defence or clotting in horseshoe crabs. Indeed, amoebocyte lysate supernatants from these animals do not display phenoloxidase activity in vitro in the same way as that of insects and crustaceans (Söderhäll et al. 1985), and there is no evidence that these animals contain prophenoloxidase in their blood systems. Accordingly, horseshoe crabs cannot be said to contain a proPO pathway, although their clotting system bears several functional and biochemical similarites to proPO (Table 2), as discussed previously by Söderhäll and Smith (1986).

Particular features which clotting in horseshoe crabs shares with proPO activation in insects and crustaceans include sensitivity to induction by LPS or β-1,3-glucans, a requirement for calcium ions and dependence upon a cascade of proteases for full activation (see reviews by Armstrong 1985, and Söderhäll and Smith 1986). The initial activator of the clotting pathway triggered by LPS is a protein termed factor C (Nakamura et al. 1986). It consists of a single chain glycoprotein with a molecular mass of 123-kDa. During activation it is converted stepwise to form a 7.9-kDa A chain and a 34-kDa B chain held together with a disulphide bond (Tokunaga et al. 1987). The 7.9-kDa A chain contains the LPS-binding site and, interestingly, is similar in sequence to mammalian complement proteins (Nakamura et al. 1986). Factor C is also very sensitive to lipid A from bacterial cell walls, or its synthetic analogues, and it is possible that lipid A might be the most effective natural trigger for clotting in these invertebrates (Nakamura et al. 1988a). A separate activator, factor B, also resides within the *Limulus* clotting system. This is sensitive to β-1,3-glucans and initiates gel formation by amoebocyte lysate extracts in a similar way to LPS (Kakinuma et al. 1981; Morita et al. 1981).

The blood of horseshoe crabs has also been shown to contain protease inhibitors (Quigley and Armstrong 1985; Nakamura et al. 1987); one of which is a 185-kDa α_2-macroglobulin secreted by the amoebocytes (Armstrong et al. 1990). In addition, the amoebocytes produce a 55-kDa protein with cell-adhesion promoting properties (Liu et al. 1991), a 15-kDa LPS-binding protein, termed the anti-LPS factor (Morita et al. 1985) and an antibacterial peptide (tachyplesin) which forms a complex with LPS and inhibits the growth of both Gram-negative and Gram-positive bacteria (Nakamura et al. 1988b). Whilst the existence of these factors does not prove that there is a proPO-type system in chelicerates, their presence emphasizes that there are certain, probably fundamental, blood proteins in the arthropod group as a whole (Table 2). Clotting, rather than phenoloxidase activity, may therefore be the common denominator.

2.4 Myriapods

Only a very small number of investigations of phenoloxidase activation have been made in other arthropod groups and these have been concerned principally with the myriapods. One of the first studies was by Bowen (1968) who reported that the blood cells of millipedes contain an enzyme which oxidizes 3-4 dihydroxyphenylalanine. In a later investigation, Krishnan and Ravindranath (1973) showed that this was phenoloxidase and that it was located in the granular haemocytes. More recently, Xylander (1992a, b) has confirmed that phenoloxidase in millipedes has a molecular mass of approximately 220-kDa and exists as an inactive proenzyme which requires proteolytic cleavage for activity. As with insects and crustaceans, enzyme activation in millipedes is calcium dependent and inhibited by PTU, but in contrast to other arthropods, activity detected by L-dopa tends to be weak and does not seem to involve a serine protease (Xylander 1992a, b; Xylander and Bogusch 1992).

Comparatively few reports dealing with the function of prophenoloxidase in myriapods have been published. One study by Nevermann and Xylander (1992) investigated encapsulation responses in centipedes and millipedes in vitro and in vivo. These workers found that a variety of foreign materials (cotton fibres, latex beads, Sephadex, and Cytodex particles) are encapsulated by the haemocytes, although the response varied according to the test material used and the species examined. The plasmatocytes and granulocytes involved in the reaction were found to be phenoloxidase-positive, with degranulation of these cells seen at the surface of implants within the first hour of the response. Deposits of melanin were subsequently observed in the capsules at 72 h (Nevermann and Xylander 1992). Whilst offering only limited information about the function of phenoloxidase in host defence in myriapods, this study indicates that the enzyme might play a role in the aggregation of haemocytes around foreign substances in a similar way to that described for insects and crustaceans. However, details of other defence responses in myriapods are lacking, so it is impossible to evaluate the extent to which prophenoloxidase participates in cell activation in these animals. In other respects, myriapods display equivalent defence capabilities to related arthropods (Table 2). For example, the haemolymph has antibacterial properties (Xylander and Nevermann 1990), but whether activity is mediated through prophenoloxidase-bearing granulocytes is unknown. Clearly, there is great scope for further investigations of the presence and biological role of phenoloxidase in this group of invertebrates.

3 Phenoloxidase Activity in Other Protostome Groups

Much less is known about the presence, activation and biological function of phenoloxidase in non-arthropod protostomes, although an increasing number of studies are now being undertaken (Table 3). In most cases, evidence for prophenoloxidase and other components in the blood of non-arthropods is anecdotal and rests with observations of melanization following wounding or tissue reactions to non-self agents. Where more concrete evidence has been obtained, it has usually been by in vitro assay of dopachrome formation by tissue extracts or by histochemical analysis of blood cell preparations. Rarely have phenoloxidase or its associated proteins been purified or characterized, and very few studies have examined the precise origin, function or regulation of putative proPO factors in vivo. Moreover, where phenoloxidase activity has been inferred from enzyme assay, confirmation that the oxidation of the substrate (usually L-dopa) was not due to peroxidase, superoxide ions or other electron acceptors has not always been presented. It is therefore difficult to assess the extent to which proPO-type factors occur in non-arthropod protostomes, the site of production of these proteins and their biological roles in the host animal.

3.1 Annelids

In annelids, melanization reactions have been noted to accompany the cellular defence reactions of the host through the development of brown bodies in the blood following infection, wounding or introduction of foreign materials (Poinar and Hess 1977; Dales 1983). The biochemical events associated with this phenomenon have not been fully investigated and phenoloxidase activity has been demonstrated in the blood cells for only a few species. Porchet-Henneré and Vernet (1992) have investigated the nature and origin of the brown pigment in capsules formed by *Nereis diversicolor* coelomocytes in vivo and established that the brown colouration is due to melanin. Enzyme cytochemistry has further revealed that phenoloxidase activity occurs in the vacuoles and Golgi apparatus of coelomocytes activated by foreign materials, and that both melanin precursors and phenoloxidase activity are located in a subpopulation of the granulocytes (Porchet-Henneré and Vernet 1992). In addition, Valembois et al. (1991) have reported that the coelomic fluid of the earthworm, *Eisenia foetida andrei* has oxidative activity towards the substrate L-dopa in vitro. The response in earthworms is blocked by subtilisin but is unaffected by trypsin and resides primarily in vesicles of the chloragocytes (Valembois et al. 1991). Importantly, the reaction in *E. foetida* is enhanced by inclusion of hydrogen peroxide and partially inhibited by catalase, indicating that the response is due to the synergistic action of both phenoloxidase and peroxidase (Valembois et al. 1991). By contrast, Smith and Söderhäll (1991) failed to detect phenoloxidase activity in the coelomic fluid

Table 3. ProPO-like activities in non-arthropod invertebrates

Activity	Acoelomates	Annelids	Molluscs	Echinoderms	Ascidians
Melanization	Yes, some sponges and coelenterates (e.g. *Antipathes fiordensis*)[a]	Yes, brown bodies formed in coelom to non-self agents (*Nereis diversicolor*)[b]	Yes, shell (several species)[c]	Yes, epidermis (*Diadema antillarum*)[d]	Yes, during non-fusion reactions between incompatible colonies (*Botryllus schlosseri*)
Phenoloxidase activity	NK	Yes, coelomocytes (*N, diversicolor*, *E. foetida*)[e]	Yes, haemocytes and mantle fluid (*Mytilus edulis*, *Biomphalaria glabrata*)[f]	Yes, coelomocytes (*D. antillarum*, *Holothuria polii* and other species)[d,g]	Yes, 'blood', usually the morula cells (*Ciona intestinalis*, *B. schlosseri* and other species)[h]
Protease activity	NK	Yes, coelomic fluid, link to proPO not shown (*E. foetida*)[i]	Yes, 'haemolymph', link to proPO not shown (*M. edulis*)[f]	Yes, coelomocytes, link to proPO not shown (*H. polii*)[k]	Yes, morula cells, may be linked to proPO activation (*C. intestinalis*)[e]
Protease inhibitor activity	NK	NK	NK	Yes, coelomic fluid, link to proPO not shown (*H. polii*)[m]	Yes, plasma, link to proPO not demonstrated, (*Halocynthia roretzi*)[n]
LPS or glucan-binding activity	NK	NK	NK	NK	Yes, 'blood', link to proPO not shown (*H. roretzi*)[o]

NK, not known. [a] Holl et al. (1992). [b] Poinar and Hess (1977), Dales (1983), Porchet-Hennére and Vernet (1992). [c] Cheng and Rifkin (1970). [d] Millott (1950, 1952), Jacobson and Millott (1953). [e] Valembois et al. (1991). [f] Bacila et al. (1961), Smith and Söderhäll (1991), Coles and Pipe (1993). [g] Canicatti and Götz (1991), Smith and Söderhäll (1991). [h] Smith and Söderhäll (1991), Smith and Peddie (1992), Jackson et al. (1993), Ballarin et al. (1993). [i] Tucková et al. (1986), Roch et al. (1991), Leipner et al. (1993). [j] Pipe (1990). [k] Canicatti (1990), Canicatti and Tschopp (1990). [l] Smith and Peddie (1992), Jackson and Smith (1993). [m] Canicatti et al. (1991). [n] Yokosawa et al. (1985). [o] Azumi et al. (1991a).

of the polychaetes, *Aphrodite aculeata* and *Arenicola marina*, as no oxidation was seen in whole blood samples incubated with L-dopa and trypsin. Whether the negative results obtained for these marine polychaetes was due to the physical and chemical conditions used in the assays is unknown. For *A. marina* and *A. aculeata*, the assays were conducted in cacodylate buffer at pH 7.0 using L-dopa as substrate, with activity measured after one hour (Smith and Söderhäll 1991). In *E. foetida*, by contrast, strongest activity was obtained at pH 9, at an optimum temperature of 37 °C and only after 8 h incubation (Valembois et al. 1991). Certainly, differences may exist in the ionic and other conditions required for oxidative activity between species and caution should be exercised in interpreting the findings where only one set of reaction conditions has been used or where more than one oxidative enzyme system is likely to be present.

An important criterion in evaluating the phylogenetic distribution of the proPO system in protostomes is whether or not phenoloxidase activity is associated with one or more serine proteases and is sensitive to activation by non-self molecules. For annelids, such information is scant and it is impossible to draw conclusions about activating cascades and roles in host defence. Several studies have established that earthworm coelomic fluid has proteolytic (Tucková et al. 1986; Leipner et al. 1993) and haemolytic activity in vitro (Roch 1979; Roch et al. 1989), and recently, Roch et al. (1991) have identified three serine proteases, A2, B2 and C2, in the coelomocytes of *E. foetida*, with molecular masses of 42, 46 and 48-kDa respectively. The activity of these proteases is pH-dependent (with two optima at 7 and 10), dose-dependent and heat-sensitive, and, while they differ in their cation requirements, they are not influenced by ionic strength (Roch et al. 1991). Attempts to enhance protease activity by prior exposure of earthworms to antigens have met with limited success (Leipner et al. 1993) and at present there is no evidence that protease activity is involved in phenoloxidase activation or that it is sensitive, directly or indirectly, to microbial carbohydrates. Protease activity in annelids has been suggested to serve a variety of roles, including nutrition and host defence (Leipner et al. 1993), but for the most part, the function of proteases in coelomic fluid remains speculative. Certainly, the haemolytic factors in earthworm coelomic fluid have been shown to have antibacterial properties (Lassegues et al. 1989), and to be linked to clotting (Valembois et al. 1988), but whether they form part of a proPO-type system is unknown.

3.2 Molluscs

With molluscs, the presence of phenoloxidase activity is somewhat clearer, although relatively few detailed analyses have been undertaken and most have concentrated on bivalves or gastropods (Table 3). The literature contains a few references to the development of brown pigmentation in haemocyte capsules formed in response to experimental infection (Cheng and Rifkin 1970), but a clear demonstration that this is due to melanin has yet to be produced. Molluscs certainly appear to possess the enzymic factors necessary for melanin synthesis as several studies have established that phenoloxidase occurs in the haemolymph of these animals. One of the first was by Bacila et al. (1961) for the freshwater snail, *Biomphalaria glabrata*. Interestingly, in this investigation, enzyme activity was detected in both albino and pigmented strains of the gastropod host showing that the genetic defect responsible for albinism in *B. glabrata* must have occurred before the formation of substrates involved in melanogenesis (Bacila et al. 1961). More significantly, visible evidence of melanization in the tissues, cuticle or test of invertebrates does not seem to be a pre-requisite for the presence of phenoloxidase activity in the blood or body fluids. Subsequently, De Aragao and Bacila (1976) partially purified phenoloxidase from *B. glabrata* by precipitation with 0.45% saturated ammonium sulphate. They found that the enzyme has low substrate specificity and a pH optimum of 7.5–8.0. It is also temperature-sensitive and requires divalent cations for activity (De Aragao and Bacila 1976). However, the precise location of the enzyme in the blood cells or plasma was not ascertained and its existence as a proenzyme not addressed. In view of the economic importance of *B. glabrata,* as an intermediate host for the parasite *Schistosoma mansonii*, and the wide distribution of this gastropod in shallow streams and ponds in tropical and subtropical regions, it is surprising that no further studies have been made on phenoloxidase activity in this animal, particularly of its biological role in host defence.

Smith and Söderhäll (1991) found that mollusc haemolymph generally displays poor oxidase activity, at least compared to levels seen in crustaceans. Weak activity was detected in the whelk, *Buccinum undatum* and the limpet, *Patella vulgata*, but in neither species of gastropod was enzyme activity enhanced by trypsin treatment (Smith and Söderhäll 1991). Similarly, the haemolymph of bivalves, specifically *Mya arenaria* and *Mytilus edulis*, was found to be almost devoid of oxidase activity, although, curiously, trypsin-inducible phenoloxidase activity was detected in the shell water from *M. edulis* (Smith and Söderhäll 1991). In bivalves, shell water contains no haemocytes and only very low levels of protein, so it is likely that the activity seen in this fluid originated elsewhere. A possible source is the mantle, a delicate structure responsible for secreting the shell, which almost invariably becomes damaged during bleeding. As yet it is unknown whether the haemocytes play a role in conveying materials (including phenoloxidase and other melanin precursors) to the mantle for tanning, but recently Coles and Pipe (1993) have obtained some cytochemical evidence that bivalves display phenoloxidase activity in the blood cells. Enzyme activity in vitro is very low and unaffected by LPS or trypsin but it appears to be enhanced by zymosan supernatant (Coles and Pipe 1993). Interestingly, *M. edulis* haemocytes were also found to stain positively for peroxidase, although phenoloxidase and peroxidase activities seem to be quite separate in this animal (Coles and Pipe 1993). Peroxidase activity is largely confined to the small granules of the eosinophilic haemocytes whereas phenoloxidase activity is restricted to the large cytoplasmic granules (Fig. 2b) (Coles and Pipe 1993). It is unclear whether phenoloxidase activity in *M. edulis* forms part of an enzyme system

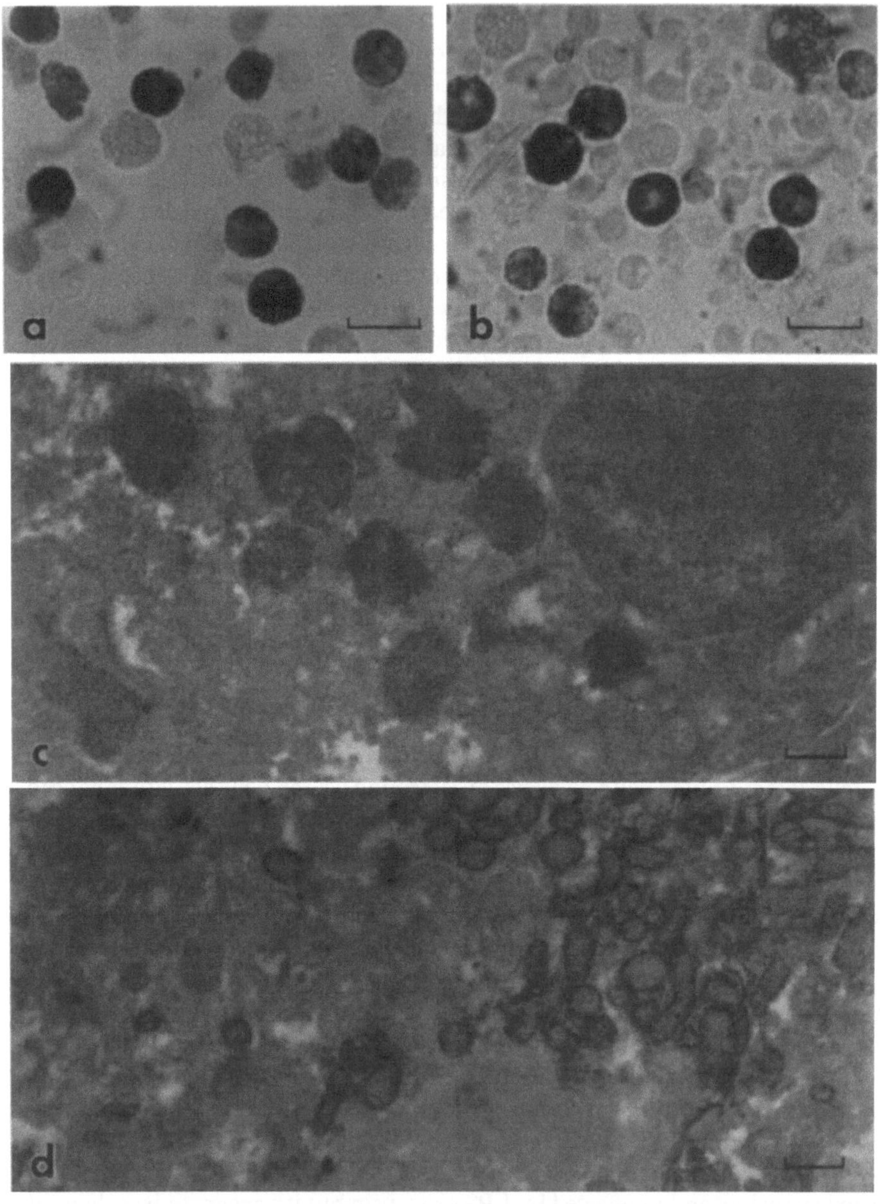

Fig. 2. Light micrographs of cytospin preparations of *Mytilus edulis* haemocytes showing granular reaction products for **a** phenoloxidase activity using L-dopa as substrate, **b** peroxidase activity using DAB as substrate. Scale *bar* = 5 µm. **c–d** Electron micrograph of *M. edulis* haemocytes showing enzyme activity as electron dense precipitate for **c** phenoloxidase localized in the large cytoplasmic granules and **d** peroxidase localized in the small cytoplasmic granules. Scale *bar* = 0.5 µm. (Coles and Pipe 1993)

sensitive to non-self molecules in a similar way to that of arthropods. Pipe (1990) has reported serine protease activity in the granulocytes of *M. edulis*, but so far no link between this and phenoloxidase activity has been demonstrated. There is also very little information about the extent to which proPO-type factors or activities might participate in host defence and there is no evidence that phenoloxidase or related factors contribute to bacterial killing, phagocytosis, agglutination, encapsulation or clotting. The question of a phenoloxidase activating system in molluscs thus remains unanswered.

3.3 Acoelomates

Very little is known about the tissue defences of acoelomates against microbial attack. Sponges appear to produce a wide range of metabolites that have antimicrobial, cytotoxic or antiviral effects, but details about the origin, biological roles and regulation of these factors have not been worked out. Little is also known about antibacterial activity in cnidarians, although it is well established that defence against overgrowth or attack by competitors is achieved through acrorhagial aggression and the firing of toxin-bearing nematocysts (Bigger 1980; Lubbock 1980). The absence of distinct circulatory systems and lack of true blood cells in acoelomate animals means that antimicrobial tissue defence must rest principally with the cells of the endoderm and ectoderm. It is therefore unlikely that a full proPO-like cascade system resides in these animals, although they may possess antimicrobial factors akin to those in other invertebrate groups.

No direct information is available about phenoloxidase activity in acoelomates but there is some indirect evidence that they are able to synthesize melanin pigments (Table 3). In a recent investigation, Holl et al. (1992) found that the ^{13}C NMR spectra of skeletal components of the New Zealand black coral, *Antipathes fiordensis* have a marked similarity to the spectra of the insect cuticle. Moreover, the predominant diphenols present in *A. fiordensis* are 3-(3, 4-dihydroxyphenyl)-DL-alanine (L-dopa) and 3, 4-dihydroxybenzaldehyde (dobal) (Holl et al. 1992), indicating that at least some of the components for tanning by quinone oxidation are present in this group. Dopamine, dopa and 5-S-cysteininyl-dopa have also been detected in the tissues of the freshwater hydrozoan, *Hydra attenuata*, and the presence and relative proportion of these compounds imply that the hydroxylation of tyrosine into dopa is mediated by a tyrosinase rather than a tyrosine hydroxylase (Carlberg 1992). Dopamine was further found to be confined to ectodermal tissue and to occur in several different cell types, including nerve cells, battery cells, nematocysts, epithelial cells and undifferentiated interstitial cells (Carlberg 1992). The abundance of dopamine in *H. attenuata* suggests that it has some biological function for the host, and because of its sporadic occurrence in neurites, it cannot serve purely as a neurotransmitter.

4 Phenoloxidase Activity in Deuterostome Invertebrates

Because of their phylogenetically strategic position close to the origin of the vertebrate line, deuterostome invertebrates have been the subject of numerous anatomical, physiological and biochemical investigations. No less with respect to their recognition and defence systems, these animals have aroused much curiosity and debate, although echinoderms and protochordates are the only groups for which a sizeable body of information has accumulated. Reports of cellular immunity in hemichordates and cephalochordates are rare and virtually nothing is known about the biochemistry of their blood cells or body fluids. The small size and/or restricted distribution of most representatives of the Larvacea, Thaliacea, Crinoidea, Ophuroidea, Cephalochordata and Hemichordata does not favour the analysis of their host defence capabilities. Instead, convenient hosts are provided by their larger and more abundant echinoid, holothurian and ascidian relatives. This review, therefore, will focus on the evidence for and against the existence of proPO-like factors or activities in these groups of deuterostome invertebrates.

4.1 Echinoderms

One of the earliest reports of the occurrence of melanin in echinoderms was provided by Fox and Scheer in 1941, but only isolated references to the presence, synthesis and biological function of the pigment have been made until recently. Studies by Jacobson and Millott (1953) and Millott (1953) have reported that the black pigment in the sea urchin, *Diadema antillarum*, and the sea cucumber, *Holothuria forskali*, is formed within the coelomic amoebocytes by an oxidative process involving phenoloxidase. In *D. antillarum*, the enzyme was found to be associated with cytoplasmic granules, and while both enzyme and substrate co-exist in the coelomocytes, melanization does not occur spontaneously in vivo (Jacobson and Millott 1953). The authors proposed that a heat-labile factor, located in the plasma, regulates the formation of melanin in the perivisceral fluid. In a much later investigation, Smith and Söderhäll (1991) confirmed that phenoloxidase activity towards L-dopa resides in the coelomocytes of *D. antillarum* and showed that this activity was enhanced by trypsin treatment. Enzyme activity was about one order of magnitude lower than that seen with equivalent samples from crustaceans, and was unaffected by LPS (Smith and Söderhäll 1991). The difference in the level of oxidase activity between these two groups of invertebrates is probably a consequence of the proportion of phenoloxidase-bearing cells in the blood samples. Unfortunately, no information is available about the number of phenoloxidase-positive cells in *D. antillarum* or other sea urchins. The effect of trypsin on phenoloxidase in *D. antillarum* indicates that the enzyme exists in echinoid coelomocytes, as in crustacean granular cells, in the form of a stable proenzyme. The lack of sensitivity of LPS, however, is less easy to interpret and more extensive investigations with other microbial carbohydrates need to be made before conclusions are made about the involvement, or not, of the enzyme in host defence. Interestingly, with whole coelomic fluid preparations (i.e. cells plus plasma) from *D. antillarum*, trypsin has no stimulatory effect on dopachrome formation in vitro (Smith and Söderhäll 1991), thus offering some indirect support for the existence of a regulatory factor in the plasma of the sea urchin as suggested by Jacobson and Millott (1953). At present, the nature of this factor, or factors, remains unknown.

Weak oxidative activities against L-dopa have also been found in the coelomocytes of the starfish, *Asterias rubens*, and the sea cucumber, *Thyone fusus* (Smith and Söderhäll 1991) (Table 3). In both species, slight enhancement in activity is achieved by pre-incubation of the cell extracts with trypsin, but only *T. fusus* cells show an increased response following LPS treatment (Smith and Söderhäll 1991). Previously, Millott (1952, 1953) had reported that phenoloxidase is present in *Thyone briareus* and *Holothuria forskali*, but gave no details about the physiological activation of the enzyme in these animals. More recently, Canicatti and Götz (1991) have examined phenoloxidase activity in the coelomic fluid of *Holothuria polii* and shown that it is confined to the coelomocytes, enhanced by chymotrypsin and blocked by serine protease inhibitors. Activity is also triggered by β-1,3-glucans and suppressed by PTU (Canicatti and Götz 1991), thereby providing some evidence that prophenoloxidase in echinoderms has certain similarities to the enzyme in arthropods.

With respect to prophenoloxidase activating proteases, Canicatti (1990) has identified peptidase activity in the coelomocytes of *H. polii* (Table 3), and shown that it is independent of cations, thermolabile and has a pH optimum of 9.5. Activity seems to be due to three serine proteases, one of which, Holozyme A, has a molecular mass of 29-kDa (Canicatti and Tschopp 1990). As yet, the relationship of these proteases to prophenoloxidase and their sensitivity to activation by microbial carbohydrates are unknown, although a heat-stable protease inhibitor has been found in *H. polii* coelomic fluid (Canicatti et al. 1991). This inhibitor is active against trypsin and subtilisin, but not chymotrypsin, and has a molecular mass of about 26-kDa (Canicatti et al. 1991). The function of this protein has not been elucidated but it has been suggested to have a protective effect for the host either by neutralizing proteases produced by microorganisms, parasites or fungi, or in regulating the activation of proteases released by the coelomocytes following non-self stimulation (Canicatti et al. 1991).

In echinoderms, the coelomocytes are known to play major roles in clotting (Canicatti and Rizzo 1991), lysis of foreign cells (Canicatti et al. 1988), antibacterial activity (Service and Wardlaw 1984; Canicatti and Roch 1989) and encapsulation (Canicatti and Quaglia 1991) (see also review by Smith 1981). Encapsulation, in particular, is a very efficient way by which echinoderms, especially holothurians, sequester foreign particles from the blood, and the development of brown pigmentation in the capsules (Hetzel 1965; Dybas and Frankboner 1986; Canicatti and D'Ancona 1989; Shinn 1989; Canicatti and Quaglia 1991), indicates that melanization reactions accompany coelomocyte reactivity in vivo. Thus, while the coelomocytes appear to produce biologically active proteins which participate in host defence (Smith 1981; Canicatti et al. 1989; Canicatti and Rizzo 1991), a clear link between cellular defence, phenoloxidase activation and the cell-derived serine proteases has not been demonstrated.

4.2 Protochordates

Amongst the protochordates, solitary and colonial ascidians have received by far the most attention with respect to their blood systems and host defence mechanisms. They are known to exhibit encapsulation reactions to implanted materials (Wright and Ermack 1982; Wright and Cooper 1983) and to display marked allorecognition tissue responses through a single, highly polymorphic gene locus (Sabbadin 1982; Rinkevitch 1992). Tissue necrosis has been observed in the contact area between the facing ampullae of incompatible colonies of botryllid ascidians (Fig. 3; Sabbadin 1982) but melanin deposition associated with the responses has seldom been mentioned in the literature. Sabbadin and co-workers have noted brown/black pigmentation in rejection reactions of *Botryllus schlosseri* colonies (Sabbadin, pers. comm.), and while cytochemical analyses indicate

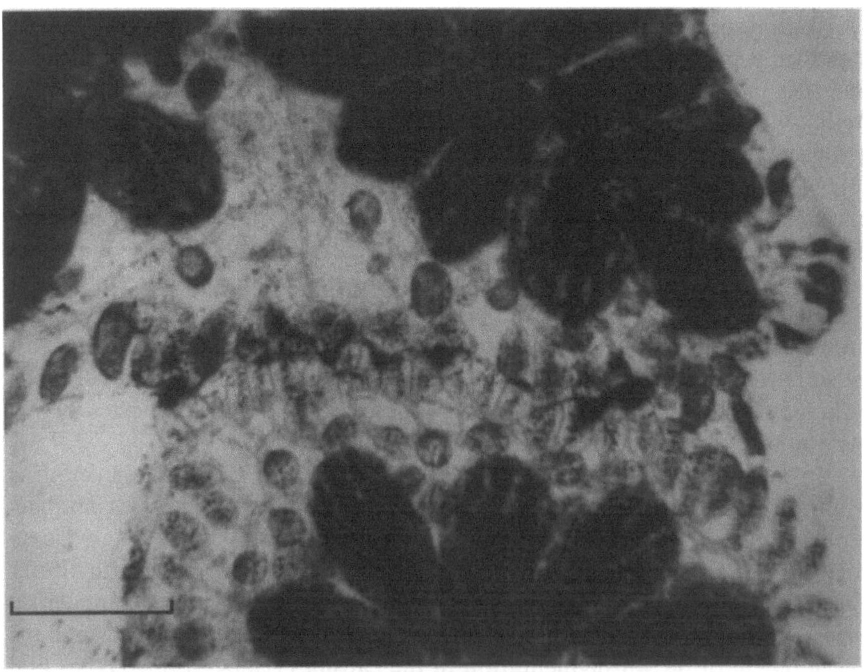

Fig. 3. Tissue necrosis in the contact area between the facing ampullae of two incompatible colonies of the colonial ascidian *Botryllus schlosseri*. Scale *bar* = 1 mm. (Sabbadin 1982)

that this is due to melanin, confirmation that phenoloxidase participates in the response has not been obtained. Other lines of enquiry, however, have offered indirect evidence that proPO-like factors or activities are present in ascidians and these may participate directly or indirectly in host defence (see below).

4.2.1 Factors Involved in Activation

Oxidative activity towards L-dopa after trypsin-treatment was found in both whole blood and blood cell lysate supernatants of the solitary sea squirt, *Ciona intestinalis* by Smith and Söderhäll (1991). Activity is also promoted by LPS or other serine type proteases, e.g. chymotrypsin and subtilisin, and resides primarily in the morula cells (Smith and Söderhäll 1991; Smith and Peddie 1992; Jackson et al. 1993). Importantly, activity in vitro is inhibited by PTU or tropolone, showing that the effect of *C. intestinalis* blood cell lysates on L-dopa is due to phenoloxidase and not peroxidase (Jackson et al. 1993). As with arthropod prophenoloxidase, phenoloxidase activity in *C. intestinalis*, is also elicited by β-1,3-glucans but not other carbohydrates (Jackson et al. 1993). In addition, the response is dose related and inhibited by benzamidine or STI, but, in contrast to insects and crustaceans, is not dependent upon divalent cations and is not suppressed by high (>100 mM) calcium concentrations (Jackson et al. 1993). Phenoloxidase activity appears to be present in at least eight other species from temperate waters, including *Ascidia mentula, Ascidia virginea, Ascidiella scabra, Ascidiella aspersa, Polycarpa pomaria, Dendrodoa grossularia, Morchellium argus* and *Botryllus schlosseri* (Jackson et al. 1993). In every case, activity appears to be enhanced by trypsin, and, where sufficient blood could be obtained for cell separation, was found to be present in the morula cells (Jackson et al. 1993). Ballarin et al. (1993) have also identified phenoloxidase activity in the morula cells of *Botryllus schlosseri* showing that the enzyme is widely distributed throughout the group and exists as a proenzyme within a distinct population of cells.

As yet, prophenoloxidase has not been purified from ascidian blood cells, so no information is available about its molecular mass or the extent of its biochemical similarity to arthropod prophenoloxidase. Likewise, little is known about the events involved in LPS or laminarin activation of prophenoloxidase in vitro. Recently, Jackson et al. (1993) have shown that phenoloxidase activity in blood cell lysate supernatants of ascidians is enhanced by exogenous serine proteases and inhibited by serine protease inhibitors, indicating that native proteases may bring about the conversion of inactive prophenoloxidase to active phenoloxidase. As yet, definitive proof that LPS-induced phenoloxidase activity in ascidians is mediated by proteases has not been presented, although Smith and Peddie (1992) and Jackson and Smith (1993) have reported LPS-sensitive protease activity in blood cell lysate supernatants of *C. intestinalis* which may be associated with the induction of prophenoloxidase in vivo. Evidence that this protease is linked to phenoloxidase activity is provided by several observations.

First, protease activity co-exists in the morula cells with prophenoloxidase (Smith and Peddie 1992). Second, protease activity is induced by a variety of lipopolysaccharides from different bacterial serotypes and the response is dose-related and inhibited by serine protease inhibitors (Jackson and Smith 1993). Third, protease activity after LPS treatment is short-lived and highest after ca. 0–15 min (Jackson and Smith 1993). Corresponding phenoloxidase activity requires at least 15 min incubation with LPS to become detectable and reaches maximum activity after ca. 30 min (Jackson and Smith 1993). Thus protease activity always precedes that of phenoloxidase and activation of prophenoloxidase only occurs once protease activity has peaked and declined (Jackson and Smith 1993). Confirmation for the role of this protease as a natural inducer of prophenoloxidase in *C. intestinalis* awaits the purification of these two factors from the blood cells and detailed examination of their interaction in vitro.

Reports of protease activity in the blood of other ascidian species are scant, so the situation for the group as a whole is difficult to assess. However, Azumi et al. (1991a) have found that the haemocytes of *Halocynthia roretzi* release a metalloprotease in response to LPS treatment in vitro. A similar effect is observed following incubation with concanavalin A, thrombin, ionomycin or phorbol myristate acetate but does not occur after treatment with β-1,3-glucans (Azumi et al. 1991a). The released protease, which has a molecular mass of 11–13 kDa, has strict specificity for succinyl-Leu-Leu-Val-Tyr-4-methylcoumaryl-7-amide substrates and is inhibited by EDTA or *o*-phenanthroline (Azumi et al. 1991a); a feature in sharp contrast to the situation for *C. intestinalis*, where protease activity has very broad substrate specificity (Jackson and Smith 1993). As yet the natural substrate and biological function for the enzyme in *H. roretzi* are unknown, and the mechanism controlling its release from the haemocytes has been only poorly documented. It is unlikely that LPS directly alters membrane fluidity of the haemocytes as protease activity in this animal is not detected upon incubation of intact cells with lipid A (Azumi et al. 1991a). Instead, the blood cells must carry a receptor for LPS (or, possibly, an LPS-plasma factor conjugate) which, following signal transduction to intracellular molecules, triggers the release of the enzyme by exocytosis (Azumi et al. 1991a). So far, there is no evidence that protease activity in *H. roretzi* is associated with phenoloxidase activity or that it plays any role in non-self recognition for the host. However, in a separate study, Azumi et al. (1991b) have identified a 12-kDa LPS-binding haemagglutinin in the haemocytes of this animal which aggregates both Gram-positive and Gram-negative bacteria in vitro. It is unclear whether this is the same protease as the one released from the cells by LPS treatment, but collectively these studies show that ascidian blood cells are capable of expressing proteases which are sensitive to microbial carbohydrates and may participate in host defence.

Regarding the regulation of protease or phenoloxidase activity in ascidian blood, Yokosawa et al. (1985) have reported the presence of trypsin inhibitor activity in the plasma of *H. roretzi*, which is heat stable and acts against a variety of trypsin-like enzymes. Purification of the factor has revealed that it is a mixture of two isoinhibitors, termed inhibitors I and II, which have molecular weights of ca. 6000 daltons and are almost identical in amino acid composition, isoelectric points and substrate specificity (Yokosawa et al. 1985). However, it does not appear to be a true α_2-macroglobulin because of its low molecular weight and insensitivity to methylamine (Yokosawa et al. 1985).

4.2.2 Functions Associated with Phenoloxidase Activity

It is disappointing that so little has been done to determine the precise function of the proteins isolated, purified and characterized from the blood of *H. roretzi* (Yokosawa et al. 1982, 1985; Harada-Azumi et al. 1987; Azumui et al. 1991a, b; Table 3) At best, there is only indirect evidence from *C. intestinalis* that proPO-like factors have a stimulatory influence on blood cell behaviour in vitro. In this animal, crude blood cell lysate supernatant preparations containing active phenoloxidase and the LPS-sensitive protease have a marked opsonic influence on the uptake of bacteria by the phagocytic amoebocytes (Smith and Peddie 1992). More importantly, opsonic potential corresponds to the degree of activity of these factors in the blood cell extracts, as inhibition of phenoloxidase and protease activity with benzamidine or STI abolishes the opsonic effect while pretreatment of the CLS with LPS produces a greatly elevated phagocytic response (Smith and Peddie 1992). Furthermore, since phenoloxidase and protease activities are confined to the morula cells and bacterial uptake is achieved only by the vacuolar and granular amoebocytes (Smith and Peddie 1992), phagocytosis in *C. intestinalis* must entail co-operation between different cell types in a manner reminiscent of that described for crustaceans and insects (see above). The finding that opsonic activity in *C. intestinalis* resides within the morula cells conflicts with earlier reports of the phenomenon in other ascidian species where enhanced uptake of erythrocytes in vitro by the phagocytes has been attributed to factors present in the plasma (Coombe et al. 1984; Kelly et al. 1992, 1993). With *C. intestinalis*, pretreatment of the test bacteria with cell-free plasma fails to stimulate their subsequent ingestion by the phagocytes (Smith and Peddie 1992), but whether plasma factors affect phagocytosis of erythrocytes is unknown. Certainly, the opsonin in *Styela clava* described by Kelly et al. (1993) may have leaked from the blood cells during sample preparation as no attempt was made to prevent clotting and cell lysis during bleeding and plasma extraction. Until the opsonin in *C. intestinalis* is purified and characterized, the extent to which it resembles the 'plasma' opsonin in *S. clava* will remain speculative.

Regarding other defence mechanisms, no clear role for phenoloxidase or the LPS-sensitive protease has been shown. However, the morula cells, which contain these factors, have been reported to participate in a number of host responses to foreign cells or materials (Table 4). For example, they migrate to and break down at sites of tunic injury (Smith 1970), participate in encapsulation reactions in the tunic (Anderson 1971; Wright and Cooper 1983), and infiltrate

Table 4. Host defence activities or factors in different invertebrate groups and their link to proPO activation

	Arthropods	Annelids	Molluscs	Echinoderms	Ascidians
Phagocytosis	Yes	Yes	Yes	Yes	Yes
Cell-derived opsonins	Yes, linked to proPO	NK	NK	NK	Yes, correlates with proPO activity
Encapsulation/nodule formation	Yes, linked to proPO	Yes, link to proPO not demonstrated	Yes, linkto proPO not demonstrated	Yes, display melanization but link to proPO not demonstrrated	Yes, link to proPO not demonstrated
Cell degranulating factors	Yes, linked to proPO	NK	NK	NK	NK
Cell adhesion factors	Yes, linked to proPO	NK	NK	NK	NK
Cell co-operation	Yes, linked to proPO	NK	NK	NK	Yes, proPO-positive morula cells provide opsonins for the phagocytes
Cell clotting	Yes, linked to proPO	Yes, link to proPO not demonstrated	Yes, link to proPO not demonstrated	Yes, link to proPO not demonstrated	Yes, link to proPO not demonstrated
Microbial killing	Yes, activity resides in proPO-bearing granular cells but link to proPO not demonstrated	Yes, link to proPO not demonstrated	Yes, link to proPO not demonstrated	Yes, link to proPO not demonstrated	Yes, proPO-positive morula cells kill Gram + and Gram – bacteria in vitro

NK, not known.

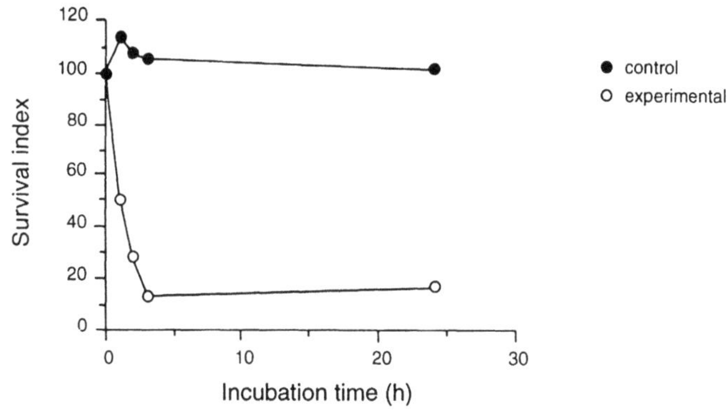

Fig. 4. Antibacterial activity of *Ciona intestinalis* blood cells against the Gram negative marine bacterium, *P. immobilis* in vitro (Smith and Hughes, unpubl.). Eight hundred microlitres of blood cell lysate supernatant were incubated with 100 µl of washed *Psychrobacter immobilis* ($1 \times 10^5\ ml^{-1}$) for 24 h at 20 °C following the methods described in Chisholm and Smith (1992). For controls, the bacteria were incubated in 800 µl of marine saline containing 0.1 % marine broth. At intervals, aliquots were removed, serially diluted and plated onto marine agar. The percentage of surviving bacteria was calculated as the survival index (Chisholm and Smith 1992). The values shown are from one representative experiment. The assay was repeated three times with similar results on each occasion, although there were variations in baseline values between runs

sites of non-fusion reactions between incompatible colonies (Tanada and Watanabe 1982). They have also been implicated in the production of antibacterial factors (Johnson and Chapman 1970; Rowley 1982; Azumi et al. 1990a), and, interestingly, Azumi et al. (1990b) have shown that five dopa-containing peptides with antibacterial properties are present in the morula cells of *H. roretzi*. As yet, direct evidence that factors released from these cells are involved in maintenance of tissue integrity has not been forthcoming. Surprisingly, the morula cells do not seem to contribute to graft rejection through cytotoxicity, as a recent study with separated blood cell populations from *C. intestinalis* has revealed that spontaneous lysis of mammalian target cells in vitro is carried out by the hyaline and phagocytic amoebocytes and not, as might be expected, by the morula cells (Peddie and Smith 1993). Instead, the morula cells must convey and deposit other biologically active factors at grafts or sites of injury. One possibility is that the morula cells carry microbicidal or disinfecting agents. Recent research in my laboratory has shown that the blood cells of *C. intestinalis* display antibacterial activity against both Gram-positive and Gram-negative bacteria in vitro (Smith and Hughes unpubl.; Fig. 4). Preliminary experiments indicate that the activity resides in the morula cells, is not due to lysozyme and persists over at least 24 h (Smith and Hughes, unpubl.). The biochemical nature of the factor(s) responsible, their mode of action and their relationship, if any, to prophenoloxidase or its associated factors is currently under investigation

5 Discussion: the proPO System in Host Defence and Immune Phylogeny

Viewing the invertebrates as a whole, it would appear that the arthropods, especially the insects and crustaceans, are the only group for which there is a substantial body of work on prophenoloxidase activation. In the last ten years good progress has been made in characterizing several of the component factors of the proPO system in these animals and in clarifying the ways they interact during host defence. This work has revealed the enormous complexity of the system and demonstrated the part played by related factors in the plasma in bringing about cell activation. Perhaps most importantly, research on proPO activation has emphasized that host defence in invertebrate animals can no longer be viewed as simply the product of a number of individual, unrelated cellular or humoral factors or activities, or even as a primitive version of vertebrate responsiveness. Instead, the proPO system has shown us that invertebrates possess sophisticated biochemical pathways in their reticuloendothelial networks and that immunity in these animals entails not only recognition and cellular reactivity but also stimulus amplification, response modulation and cell–cell signalling. Furthermore, as our understanding of proPO activation is still far from complete, new factors associated with the system, the nature and biological importance of which we are presently unaware, may yet be found in arthropod blood.

As far as non-arthropod invertebrates are concerned, it is clear that melanin and/or melanin precursors occur in several phyla, including both protostomes and deuterostomes (Table 3). Less certain is whether these factors form part of a cascade of biologically active proteases and related molecules which function in host defence. Phenoloxidase activity has been identified in the blood of annelids, molluscs, echinoderms and ascidians, but has been shown to be present as a stable proenzyme sensitive to microbial carbohydrates in only one bivalve (*Mytilus edulis*), one holothurian (*H. polii*) and a few species of ascidians (Table 3). Ascidians are also the only non-arthropod group for which there is some evidence, albeit indirect, that LPS or glucan-induced phenoloxidase activity is associated with one or more serine proteases (Table 4). The presence of an LPS-binding haemagglutinin, protease inhibitors and cell-derived opsonins in these animals provide further intriguing clues that they share certain bioactive factors of activities with the arthropods (Table 5). Thus, ascidians offer the best prospect for finding a proPO-type system outside of the arthropods. The work on molluscs, annelids and echinoderms shows encouraging trends and warrants further research, especially with respect to the mechanisms underlying cell activation, but insufficient data have accumulated to draw conclusions about the presence or absence of proPO-type pathways in their blood systems.

At first it may seem surprising that two such taxonomically separate invertebrate groups as the ascidians and arthropods should both display proPO-related factors or activities in their blood systems. However, evaluation of

Table 5. Comparison of host defence capabilities in crustaceans and ascidians

Crustaceans	Ascidians
Rely on infammatory-type responses (e.g. phagcytosis, nodule/capsule formation, production of antimicrobial agents) to host defence	Rely on inflammatory-type responses for host defence
Phagocytosis and nodule/capsule formation entails co-operation between different blood cell types	Phagocytosis entails co-operation between different blood cell types
Phenoloxidase and protease activities contained within granular and semi-granular blood cells	Phenoloxidase and protease activities contained within morula cells
Granular cells contain opsonins	Morula cells contain opsonins
Granular cells contain antibacterial agents	Morula cells contain antibacterial agents
Protease inhibitors present in plasma, probably regulate proPO activation	Protease inhibitors present in plasma, function unknown
Glucan and LPS binding proteins present in blood cells or plasma	LPS-binding protein present in blood cells

For references, see text.

phylogenetic trends in immunity based on comparative studies of extant species depends upon how we view the evolutionary relationships between various animal types. Numerous versions of phylogenetic trees have been put forward; some based on classical anatomical and embryological studies, others on more recent molecular findings. Differences between them have bearing on the way we interpret the evolution of immune capability within the animal kingdom. Thus, while it is improbable that proPO factors from arthropods should exhibit any homology with equivalent factors in ascidians from a monophyletic standpoint, parallel evolution of the pathway is more likely from a polyphyletic stance.

Importantly too, the evaluation of the proPO system in a range of protostome and deuterostome invertebrates, as described in this chapter, rests largely upon the notion that prophenoloxidase is the definitive marker for the system in different animal species. This, of course, may not be the case. Chelicerates, for example, do not display phenoloxidase activity in their body fluids yet their clotting system has many similarities to the proPO pathway of insects and crustaceans (Table 1). Clearly, the enzyme is not a pre-requisite for the existence of a parallel system in invertebrates. Indeed, it may simply represent the end point of a pathway which finds full expression in only a small number of animal groups. Other molecules might be more appropriate indicators of a shared system, but precisely which markers constitute the best alternative is difficult to judge at this stage.

Some pointers might be found amongst other factors related to proPO activation, especially if they show functional overlap across the different invertebrate groups. Protease activity, for example, has been reported to occur in the blood of many species, including earthworms, insects, crustaceans, horseshoe crabs, bivalves, holothurians and ascidians (Tables 2 and 3). While some of these enzymes participate in clotting, proPO activation and/or antimicrobial activity, others may serve important roles in nutrition, reproduction or development. At present, there is no indication that protease activity is primarily associated with host defence and it is likely that protease enzymes arose independently and without the need for immunological reactivity. Similarly, protease inhibitors probably comprise a heterogeneous collection of regulatory molecules, although some, at least, might assist host protection by blocking the action of aggressive enzymes secreted by parasites or pathogens. More suitable candidates for primitive defence molecules are the LPS- and glucan-binding proteins. The ability of these proteins to bind quasi-specifically to microbial carbohydrates makes them central to all defence phenomena. Whatever the nature of the subsequent biochemical or immunological events, e.g. microbial killing, opsonization, clotting, exocytosis or proPO activation, an early step must be the interaction between the host proteins and the surface of the foreign material. So far, evidence that glucan or LPS-binding molecules influence host defence comes mainly from crayfish, horseshoe crabs and, to a lesser extent, ascidians. Glucan or LPS-binding proteins are also present in insects, although their biological effects still need to be clarified, and as yet equivalent factors have not been identified in other invertebrate groups. Since processes, such as antibacterial activity and opsonization which depend upon direct contact between host factors and microbial cells, occur universally throughout the Invertebrata, it is possible that glucan or LPS-binding factors might be found in several invertebrate phyla.

6 Conclusions and Future Perspectives

To conclude, this bottom-up survey of proPO-type factors or activities in the various invertebrate groups has shown that while phenoloxidase activity occurs in many groups outside of the insects and crustaceans, only ascidians show evidence of a proPO-like system (as defined by the presence of prophenoloxidase and associated protease). There are, however, several factors or activities, which resemble components of the pathway in arthropods, and which are present in a range of protostome and deuterostome animals. Some appear to contribute directly to host defence, others may secondarily take on an immunological function, but we do not yet know the degree of relatedness between these factors in different groups. Neither is it clear whether these agents collectively form part of a recognition or defence pathway shared in total or in part by a number of phyla. The paucity of information therefore prevents this bottom-up analysis from establishing the proPO system as a common defence pathway for invertebrate animals. However, it has highlighted the importance of lesser known molecules, such as the glucan or LPS-binding molecules, as possible contenders for unifying factors.

Future work on the proPO system is still highly merited. In arthropods, much remains to be learnt about the biochemical nature of the constituent factors and the mechanism(s) by which they exert their biological effects. In particular, the relationship of proPO-activating proteins to bacterial killing and the nature of the cell-derived opsonins needs to be addressed. With molluscs and annelids, information is wanting about the sensitivity of their blood proteins to LPS or other microbially derived carbohydrates, the nature of any LPS or glucan-binding proteins within the blood and whether these animals contain protease inhibitors which regulate immune reactivity. For ascidians, information is needed about the biochemical characteristics (e.g. molecular mass, subunit structure and amino acid sequence) of the putative proPO factors and of the relationship between the LPS-binding haemagglutinin, proPO activation, antibacterial activity and opsonization. From such studies, it might be possible for us to elucidate the extent of homology between the various proPO factors in different invertebrate groups and to pinpoint those molecules which have been conserved during evolutionary development.

Acknowledgements. I would like to thank Dr. Richard Pipe and Ms Jackie Coles (Plymouth Marine Laboratory, England), Prof. Armando Sabbadin (University of Padova, Italy), Dr. June Chisholm (Ninewells Hospital, Dundee, Scotland) and Dr. Karen Bell and Miss Shirley Hughes (St. Andrews University) for allowing me to use unpublished information and for providing the figures used in this review. The original work described in this chapter was supported by grants from the Science and Engineering Research Council, the Natural Environment Research Council and the Nuffield Foundation, UK.

References

Anderson RS (1971) Cellular responses to foreign bodies in the tunicate *Molgula manhattensis*. Biol Bull 141: 91–98

Andersson K, Sun SC, Boman HG, Steiner H (1989) Purification of the prophenoloxidase from *Hyalophora cecropia* and four proteins involved in its activation. Insect Biochem 19: 629–637

Anggraeni T, Ratcliffe NA (1991) Studies on cell–cell co-operation during phagocytosis by purified haemocyte populations of the wax moth *Galleria mellonella*. J Insect Physiol 37: 453–460

Armstrong PB (1985) Amoebocytes of the american horseshoe crab, *Limulus*. In: Cohen WD (ed) *Blood cells of marine invertebrates*. Alan R Liss, New York, pp 253–258

Armstrong PB, Quigley, JP, Rickles, FR (1990) The *Limulus* blood cell secretes α_2-macroglobulin when activated. Biol Bull 178: 137–143

Ashida M (1971) Purification and characterization of prephenoloxidase from haemolymph of the silkworm, *Bombyx mori*. Arch Biochem Biophys 144: 749–762

Ashida M (1981) A cane sugar factor suppressing activation of prophenoloxidase in hemolymph of the silkworm, *Bombyx mori*. Insect Biochem 11: 57–65

Ashida M, Dohke K (1980) Activation of prophenoloxidase by the activating enzyme of the silkworm, *Bombyx mori*. Insect Biochem 10: 37–47

Ashida M, Söderhäll K (1984) The prophenoloxidase activating system in crayfish. Comp Biochem Physiol 77B: 21–26

Ashida M, Yoshida H (1988) Limited proteolysis of prophenoloxidase during activation by microbial products in insect plasma and effect of phenoloxidase on electrophoretic mobilities of plasma proteins. Insect Biochem 18: 11–19

Ashida M, Ishizaki Y, Iwahama M (1983) Activation of prophenoloxidase by bacterial cell walls or β-1,3-glucans in plasma of the silkworm, *Bombyx mori*. Biochem Biophys Res Commun 117: 562–568

Aso Y, Kramer KJ, Hopkins TL, Lookhart GL (1985) Characterization of haemolymph protyrosinase and a cuticular activator from *Manduca sexta* (L). Insect Biochem 15: 9–17

Aspán A, Söderhäll K (1991) Purification of prophenoloxidase from crayfish blood cells and its activation by an endogenous serine protease. Insect Biochem 21: 363–373

Aspán A, Sturtevant JE, Smith VJ, Söderhäll K (1990) Purification and characterization of a prophenoloxidase activating enzyme from crayfish blood cells. Insect Biochem 20: 709–718

Azumi K, Yokosawa H, Ishii S-I (1990a) Halocyamines: novel antimicrobial tetrapeptides-like substances isolated from the hemocytes of the solitary ascidian, *Halocynthia roretzi*. Biochemistry 29: 159–165

Azumi K, Yokosawa H, Ishii S-I (1990b) Presence of 3,4 dihydroxyphenylalanine-containing peptides in hemocytes of the ascidian, *Halocynthia roretzi* Experientia 46: 1020–1023

Azumi K, Yokosawa H, Ishii S-I (1991a) Lipopolysaccharide induces release of a metalloprotease from hemocytes of the ascidian, *Halocynthia roretzi*. Dev Comp Immunol 15: 1–7

Azumi K, Ozeki S, Yokosawa H, Ishii S-I (1991b) A novel lipopolysaccharide-binding hemagglutinin isolated from hemocytes of the solitary ascidian, *Halocynthia roretzi*: it can agglutinate bacteria. Dev Comp Immunol 15: 9–16

Bacila M, Duarte JH, Voss DO (1961) Purificaçao da dihidroxifeniloxidae de planorbidideos negros e albinos e estudo espectrosopico dos productos de reacao. An Acad Bras Cienc 33: xiii

Ballarin L, Cima F, Sabbadin A (1993) Histochemical staining and characterization of the colonial ascidian *Botryllus schlosseri* hemocytes. Boll Zool 60: 19–24

Barracco MA, Duvic B, Söderhäll K (1991) The β-1,3-glucan binding protein from the crayfish *Pacifastacus leniusculus* when reacted with a β-1,3-glucan induces spreading and degranulation of crayfish granular cells. Cell Tissue Res 266: 4910–4917

Bigger CH (1980) Interspecific and intraspecific acrorhagial aggressive behavior among sea anemones. A recognition of self and non-self. Biol Bull 159: 117–134

Boman HG, (1991) Antibacterial peptides: key components needed in immunity. Cell 65: 205–207

Boman HG, Hultmark D (1987) Cell-free immunity in insects. Annu Rev Microbiol 41: 103–126

Bowen RC (1968) Tyrosinase in millipede haemocytes. Trans Am Microsc Soc 87: 390–392

Brookman JL, Rowley AF, Ratcliffe NA (1989) Studies on nodule formation in locusts following injection of microbial products. J Invertebr Pathol 53: 315–323

Canicatti C (1990) Protease activity in *Holothuria polii* coelomic fluid and coelomocyte lysate. Comp Biochem Physiol 95B: 145–148

Canicatti C, D'Ancona G (1989) Cellular aspects of *Holothuria polii* immune response. J Invertebr Pathol 53: 152–158

Canicatti C, Götz P (1991) Dopa oxidation by *Holothuria polii* coelomocyte lysate. J Invertebr Pathol 58: 305–310

Canicatti C, Quaglia A (1991) Ultrastructure of *Holothuria polii* encapsulating body. J Zool Lond 224: 419–429

Canicatti C, Rizzo A (1991) A 220 kDa coelomocyte aggregating factor involved in *Holothuria polii* cellular clotting. Eur J Cell Biol 56: 79–83

Canicatti C, Roch Ph (1989) Studies on *Holothuria polii* (Echinodermata) antibacterial proteins. I. Evidence and activity of coelomocyte lysozyme. Experientia 45: 756–759

Canicatti C, Tschopp J (1990) Holozyme A: one of the serine proteases of *Holothuria polii* coelomocytes. Comp Biochem Physiol 96B: 739–742

Canicatti C, Ciulla D, Farina-Lipari E (1988) The hemolysin producer hemocytes in *Holothuria polii*. Dev Comp Immunol 12: 729–736

Canicatti C, Miglietta A, Cooper EL (1989) In vitro release of biologically active molecules during the clotting reaction in *Holothuria polii*. Comp Biochem Physiol 94A: 483–488

Canicatti C, Roch Ph, Stabili L, Pagliara P (1991) Protease inhibitor in *Holothuria polii* coelomic fluid. Comp Biochem Physiol 98B: 593–596

Carlberg M (1992) Localization of dopamine in the freshwater hydrozoan, *Hydra attenuata*. Cell Tissue Res 270: 601–607
Chen JS, Rolle RS, Marshall MR, Wei CI (1991) Comparison of phenoloxidase activity from Florida spiny lobster and Western Australian lobster. *J Food Sci* 56: 154–160
Cheng TC, Rifkin E (1970) Cellular reactions in marine molluscs in response to helminth parasitism. In: Snieszko SF (ed) *A symposium on diseases of fishes and shellfishes*. Am Fish Soc Washington, DC, pp 443–496
Chisholm JRS (1993) Antibacterial activity in the blood cells of *Carcinus maenas* (L) and other crustaceans. PhD Thesis, Univ St Andrews, Scotland
Chisholm JRS, Smith VJ (1992) Antibacterial activity in the haemocytes of the shore crab, *Carcinus maenas*. J Mar Biol Assoc UK 72: 529–542
Chisholm JRS, Smith VJ (1994) Variation of antibacterial activity in the haemocytes of the shore crab, *Carcinus maenas* with temperature. J Mar Biol Assoc UK 74: 979–982
Coles JA, Pipe RK (1993) Phenoloxidase activity in the haemolymph and haemocytes of the marine mussel, *Mytilus edulis*. Fish Shellfish Immunol 4: 337–352
Coombe DR, Ely PL, Jenkin CR (1984) Particle recognition by haemocytes from the colonial ascidian *Botrylloides leachii* evidence that the *B. leachii* HA-2 agglutinin is opsonic. J Comp Physiol 154b: 509–521
Dales RP (1983) Observations on granulomata in the polychaetus annelid *Nereis diversicolor*. J Invertebr Pathol 42: 288–291
De Aragao GA, Bacila M (1976) Purification and properties of a polyphenoloxidase from the freshwater snail, *Biomphalaria glabrata*. Comp Biochem Physiol 54B: 179–182
Dohke K (1973) Studies on prephenoloxidase-activating enzyme from cuticle of the silkworm, *Bombyx mori*. Arch Biochem Biophys 157: 203–209
Dularay B, Lackie AM (1985) Haemocyte encapsulation and the prophenoloxidase-activating pathway in the locust *Schistocerca gregaria* Forsk. Insect Biochem 15: 827–834
Dunphy GB (1991) Phenoloxidase activity in the serum of two species of insects, the gypsy moth, *Lymantria dispar* (Lymantriidae) and the greater wax moth, *Galleria mellonella* (Pyralidae). Comp Biochem Physiol 98B: 535–538
Durliat M (1985) Clotting processes in Crustacea Decapoda. Biol Rev 60: 473–498
Duvic B, Söderhäll K (1990) Purification and characterization of a β-1,3-glucan binding protein from plasma of the crayfish *Pacifastacus leniusculus*. J Biol Chem 265: 9327–9332
Dybas L, Frankboner PV (1986) *Holothuria* survival strategies: mechanisms for the maintenance of a bacteriostatic environment in the coelomic cavity of the sea cucumber, *Parastichopus californicus*. Dev Comp Immunol 10: 311–330
Eguchi M, Shomoto K (1985) Purification and properties of a chymotrypsin inhibitor from the silkworm haemolymph. Comp Biochem Physiol 81B: 301–307
Fox DL, Scheer BT (1941) Comparative studies of the pigments of some Pacific coast echinoderms. Biol Bull 80: 441–455
Fuller GM, Doolittle RF (1971a) Studies of invertebrate fibrinogen. I. Purification and characterization of fibrinogen from the spiny lobster. Biochemistry 10: 1305–1311
Fuller GM, Doolittle RF (1971b) Studies of invertebrate fibrinogen. II. Transformation of lobster fibrinogen into fibrin. Biochemistry 10: 1311–1315
Häll L, Söderhäll K (1982) Purification and properties of a protease inhibitor from crayfish hemolymph. J Invertebr Pathol 39: 29–37
Hall M, Söderhäll K, Sottrup-Jensen L (1989) Amino acid sequence around the thioester of α_2-macroglobulin from plasma of the crayfish *Pacifastacus leniusculus*. FEBS Lett 254: 111–114
Harada-Azumi KH, Yokosawa H, Ishii S-I (1987) *N*-Acetyl galactosamine specific lectin, a novel lectin in the haemolymph of the ascidian *Halocynthia roretzi*: isolation, characterization and comparison with the galactose-specific lectin. Comp Biochem Physiol 88B: 375–381
Hergenhahn H-G, Aspán A, Söderhäll K (1987) Purification and characterization of a high *M*r proteinase inhibitor of prophenoloxidase activation from crayfish plasma. Biochem J 248: 223–228

Hergenhahn H-G, Hall M, Söderhäll K (1988) Purification and characterization of an α_2-macroglobulin-like proteinase inhibitor from plasma of crayfish *Pacifastacus leniusculus*. Biochem J 255: 801–806
Hetzel HR (1965) Studies on holothurian coelomocytes. The origin of coelomocytes and formation of brown bodies. Biol Bull 128: 102–111
Holl SM, Scaefer J, Goldberg WM, Kramer KJ, Morgan TD, Hopkins TL (1992) Comparison of black coral skeleton and insect cuticle by a combination of carbon-^{13}NMR and chemical analysis. Arch Biochem Biophys 292: 107–111
Iwama R, Ashida M (1986) Biosynthesis of prophenoloxidase in hemocytes of larval hemolymph of the silkworm *Bombyx mori*. Insect Biochem 16: 547–555
Jackson AD, Smith VJ (1993) LPS-sensitive protease activity in the blood cells of the solitary ascidian, *Ciona intestinalis* (L). Comp Biochem Physiol 106B: 505–512
Jackson AD, Smith VJ, Peddie CM (1993) In vitro phenoloxidase activity in the blood of *Ciona intestinalis* and other ascidians. Dev Comp Immunol 17: 97–108
Jacobson FW, Millott N (1953) Phenolases and melanogenesis in the coelomic fluid of the echinoid, *Diadema antillarum* Phillippi. Proc R Soc 141B: 231–246
Janeway CA (1989) A primitive immune system. Nature (Lond) 341: 108
Jarosz J (1993) Induction kinetics of immune antibacterial proteins in pupae of *Galleria mellonella* and *Pieris brassicae*. Comp Biochem Physiol 106B: 415–421
Johansson MW, Söderhäll K (1988) Isolation and purification of a cell adhesion factor from crayfish blood cells. J Cell Biol 106: 1795–1803
Johansson MW, Söderhäll K (1989a) Cellular immunity in crustaceans and the proPO system. Parasitol Today 5: 171–176
Johansson MW, Söderhäll K (1989b) A cell adhesion factor from crayfish hemocytes has degranulating activity towards crayfish granular cells. Insect Biochem 19: 183–190
Johansson MW, Söderhäll K (1989c) A peptide containing the cell adhesion sequence RGD can mediate degranulation and cell adhesion of crayfish granular haemocytes in vitro. Insect Biochem 19: 573–579
Johnson PT, Chapman FA (1970) Comparative studies on the in vitro response of bacteria to invertebrate body fluids. II. *Aplysia california* and *Ciona intestinalis*. J Invertebr Pathol 16: 259–267
Jomori T, Kubo T, Natori S (1990) Purification and characterization of a lipopolysaccharide-binding protein from hemolymph of the American cockroach, *Periplaneta americana*. Eur J Biochem 190: 201–206
Kakinuma A, Asano T, Torrii H, Sugino Y (1981) Gelation of *Limulus* amoebocyte lysate by an antitumor (1–3) 1-β-D-glucan. Biochem Biophys Res Commun 101: 434–439
Kang S-H, Fuchs MS (1980) Purification and partial characterization of a protease inhibitor from *Drosophila melanogaster*. Biochem Biophys Acta 611: 379–383
Kelly KL, Cooper EL, Raftos DA (1992) Purification and characterization of a humoral opsonin from *Styela clava*. Comp Biochem Physiol 103B: 749–753
Kelly KL, Cooper EL, Raftos DA (1993) A humoral opsonin from the solitary urochordate, *Styela clava*. Dev Comp Immunol 17: 29–39
Klein J (1989) Are invertebrates capable of anticipatory immune responses? Scand J Immunol 29: 499–505
Kopacek P, Hall M, Söderhäll K (1993) Characterization of a clotting protein, isolated from plasma of the freshwater crayfish, *Pacifastacus leniusculus*. Eur J Biochem 213: 591–597
Krishnan G, Ravindranath MH (1973) Blood cell phenoloxidase of millipedes. J Insect Physiol 19: 647–653
Lassegues M, Roch Ph, Valembois P (1989) Antibacterial activity of *Eisenia fetida andrei* coelomic fluid: evidence, induction and animal protection. J Invertebr Pathol 53: 1–6
Leipner C, Tucková L, Rejnek J, Langner J (1993) Serine proteases in coelomic fluids of annelids *Eisenia foetida* and *Lumbricus terrestris*. Comp Biochem Physiol 105B: 637–641
Leonard C, Söderhäll K, Ratcliffe NA (1985a) Studies on prophenoloxidase and protease activity of *Blaberus craniifer* haemocytes. Insect Biochem 15: 803–810
Leonard C, Ratcliffe NA, Rowley AF (1985b). The role of prophenoloxidase activation in non-self recognition and phagocytosis by insect blood cells. J Insect Physiol 31: 789–799

Liang Z, Linblad P, Beavais A, Johansson MW, Latğe J-P, Hall M, Cerenius L, Söderhäll K (1992) Crayfish α_2-macroglobulin and 76 kDa protein; their biosynthesis and subcellular localization of the 76 kDa protein. J Insect Physiol 38: 987–995

Liu T, Lin Y, Cislo T, Minetti CASA, Baba JMK, Liu T-Y (1991) Limunectin: a phosphocholine binding protein from *Limulus* with adhesion promoting properties J Biol Chem 266: 14813–14821

Lubbock R (1980) Clone specific cellular recognition in a sea anemone. Proc Natl Acad Sci USA 77: 6667–6669

Millott N (1950) Integumentary pigmentation and the coelomic fluid of *Thyone briarus* (Lesueur). Biol Bull 99: 343–344

Millott N (1952) The occurrence of melanin and phenolases in *Holothuria forskali* (Delle Chiaje). Experientia 8: 301–302

Millott N (1953) Observations on the skin pigment and amoebocytes, and occurrence of phenolases in the coelomic fluid of *Holothuria forkali* Delle Chiarje. J Mar Biol Assoc UK 31: 529–539

Morita TS, Tanaka T, Nakamura T, Iwanaga S (1981) A new (1–3)-β-D-glucan mediated coagulation pathway found in *Limulus* amoebocytes. FEBS Lett 129: 318–321

Morita T, Ohtsubo S, Nakamura T, Tanaka S, Iwanaga S, Ohashi K, Niwa M (1985) Isolation and biological activities of *Limulus* anticoagulant (anti-LPS factor) which interacts with lipopolysaccharide (LPS). J Biol Chem 97: 1611–1620

Munn EA, Bufton SF (1973) Purification and properties of a phenoloxidase from the blowfly *Calliphora erythrocephala*. Eur J Biochem 35: 3–10

Nakamura T, Morita T, Iwanaga S (1986) Lipopolysaccharide-sensitive protease zymogen (factor C) found in *Limulus* hemocytes. Isolation and characterization. Eur J Biochem 154: 511–521

Nakamura T, Hirai T, Tokunaga F, Kawabata S, Iwanaga S (1987) Purification and amino acid sequence of Kunitz-type protease inhibitor found in the hemocytes of the horseshoe crab, (*Tachypleus tridentatus*). J Biol Chem 101: 1279–1306

Nakamura T, Tokunaga F, Morita T, Iwanaga S, Kusumoto S, Shiba T, Kobayashi T, Inoue K (1988a) Intracellular serine-protease zymogen, factor C, from horseshoe crab hemocytes. Its activation by synthetic lipid A analogues and acidic phospholipids. Eur J Biochem 176: 89–94

Nakamura T, Furunaka H, Miyata T, Tokunaga F, Muta T, Iwanaga S, Niwa M, Takao T, Shimonishi Y, (1988b) Tachyplesin, a class of antimicrobial peptide from the hemocytes of the horseshoe crab (*Tachypleus tridentatus*). J Biol Chem 263: 16709–16713

Nappi AJ, Vass E (1993) Melanogenesis and the generation of cytotoxic molecules during insect cuticular immune reactions. Pigment Cell Res 6: 117–126

Naqvi SNH, Karlson P (1979) Purification of prophenoloxidase in the haemolymph of *Calliphora vicinia* (R & D). Arch Int Physiol Biochem 87: 687–695

Nevermann L, Xylander WER (1992) Cellular immune responses in centipedes and millipedes (Arthropoda, Tracheata). 25th Annu Meet Soc Invertebr Pathol, Heidelberg, Germany, Aug 16–21 (Absrt)

Ochiai M, Ashida M (1988) Purification of a β-1, 3-glucan recognition protein in the prophenoloxidase activating system from hemolymph of the silkworm, *Bombyx mori*. J Biol Chem 263: 12056–12062

Pawelek JM, Lerner AB (1978) 5, 6-Dihydroxyindol is a melanin precursos showing potent-cytotoxicity. Nature (Lond) 276: 627–628

Peddie CM, Smith VJ (1993) In vitro spontaneous cytotoxic activity against mammalian target cells by the hemocytes of the solitary ascidian, *Ciona intestinalis*. J Exp Zool 267: 616–623

Persson M, Vey A, Söderhäll K (1987) Encapsulation of foreign particles in vitro by separated blood cells from crayfish, *Astacus leptodactylus*. Cell Tissue Res 247: 409–415

Pipe RK (1990) Hydrolytic enzymes associated with the granular haemocytes of the marine mollusc, *Mytilus edulis*. Histochem J 22: 595–603

Poinar GO Jr, Hess RT (1977) Immune responses in the earthworm *Aporrectodea trapezoides* (Annelida) against *Rhabditis pellio* (Nematoda). In: Bulla LA, Cheng TC, (eds) *Comparative pathobiology*, vol 3. Plenum Press, New York, pp 69–84

Porchet-Henneré E, Vernet G (1992) Cellular immunity in an annelid (*Nereis diversicolor*, Polychaeta): production of melanin by a sub-population of granulocytes. Cell Tissue Res 269: 167–174

Pye AE (1974) Microbial activation of prophenoloxidase from immune insect larvae. Nature (Lond) 251: 610–613

Quigley JP, Armstrong PB (1985) A homologue of α_2-macroglobulin purified from the hemolymph of the horseshoe crab, *Limulus polyphemus*. J Biol Chem 258: 7903–7906

Rantamäki J, Durrant H, Liang Z, Ratcliffe NA, Duvic B, Söderhäll K (1991) Isolation of a 90 kDa protein from haemocytes of *Blaberus craniifer* which has similar functional and immunological properties to the 76 kDa protein from crayfish haemocytes. J Insect Physiol 37: 627–634

Ratcliffe NA, Leonard CM, Rowley AF (1984) Prophenoloxidase activation: non-self recognition and cell co-operation in insect immunity. Science 226: 557–559

Ratcliffe NA, Rowley AF, Fitzgerald SW, Rhodes CP (1985) Invertebrate immunity: basic concepts and recent advances. Int Rev Cytol 97: 183–350

Rinkevitch B (1992) Aspects of the incompatibility nature in botryllid ascidians. Anim Biol 1: 17–28

Roch Ph (1979) Protein analysis of earthworm coelomic fluid: I. Polymorphic system of the natural hemolysin of *Eisenia fetida andrei*. Dev Comp Immunol 3: 599–608

Roch Ph, Canicatti C, Valembois P (1989) Interactions between earthworm hemolysins and sheep red blood cell membranes. Biochim Biophys Acta 983: 193–198

Roch Ph, Stabili L, Pagliara P (1991) Purification of three serine proteases from the coelomic cells of earthworms (*Eisenia fetida*). Comp Biochem Physiol 98B: 579–602

Rolle, RS, Marshall MR, Wei CI, Chen JS (1990) Phenoloxidase forms of the Florida spiny lobster; immunological and spectropolarimetric characterization. Comp Biochem Physiol 97B: 483–489

Rowley AF (1982) The blood cells of *Ciona intestinalis*: an electron probe X-ray study. J Mar Biol Assoc UK 62: 607–620

Sabbadin A (1982) Formal genetics of ascidians. Am Zool 22: 765–773

Sasaki T (1984) Amino acid sequence of novel Kunitz-type chymotrypsin inhibitor from hemolymph of silkworm larvae, *Bombyx mori*. FEBS Lett 168: 227–230

Saul S, Sugumaran M (1986) Protease inhibitor controls prophenoloxidase activation in *Manduca sexta*. FEBS Lett 208: 113–116

Saul S, Sugumaran M (1987) Protease mediated prophenoloxidase activation in the tobacco hornworm, *Manduca sexta*. Arch Insect Biochem Physiol 5: 1–11

Saul S, Sugumaran M (1988) Prophenoloxidase activation in the hemolymph of *Sarcophaga bullata* larvae. Arch Insect Biochem Physiol 7: 91–103

Service M, Wardlaw AC (1984). Echinochrome A as a bactericidal substance in the coelomic fluid of *Echinus esculentus* (L). Comp Biochem Physiol 79B: 161–165

Shinn GL (1989) Reproduction of *Anoplodium hymanae*, a turbellarian flatworm (Neorhabdocoela, Umagillidae) inhabiting the coelom of sea cucumbers; production of egg capsules and escape of infective stages without evisceration of the host. Biol Bull 169: 182–198

Simpson BK, Marshall MR, Otwell WS (1988) Phenoloxidase from pink and white shrimp; kinetic and other properties. J Food Biochem 12: 205–209

Smith MJ (1970) The blood cells and tunic of the ascidian *Halocynthia aurantium* (Pallas). I. Haematology, tunic morphology and partition of the cells between blood and tunic. Biol Bull 138: 354–378

Smith VJ (1981) The echinoderms. In: Ratcliffe NA, Rowley AF (eds) Invertebrate blood cells, vol II. Academic Press, London, pp 513–562

Smith VJ (1991) Invertebrate immunology: phylogenetic, ecotoxicological and biomedical implications. Comp Haematol Int 1: 61–76

Smith VJ, Peddie CM (1992) Cell co-operation during host defence in the solitary tunicate, *Ciona intestinalis* (L). Biol Bull 183: 211–219

Smith VJ, Söderhäll K (1983a) β 1,3-Glucan activation of crustacean hemocytes in vitro and in vivo. Biol Bull 164: 299–314

Smith VJ, Söderhäll K (1983b) Induction of degranulation and lysis of haemocytes in the freshwater crayfish *Astacus astacus* by components of the prophenoloxidase activating system in vitro. Cell Tissue Res 233: 295–303
Smith VJ, Söderhäll K (1986) Cellular immune mechanisms in the Crustacea. Symp Zool Soc Lond 56: 59–79
Smith VJ, Söderhäll K (1991) A comparison of phenoloxidase activity in the blood of marine invertebrates. Dev Comp Immunol 15: 251–261
Smith VJ, Söderhäll K, Hamilton M (1984) β-1, 3-glucan induced cellular defence reactions in the shore crab, *Carcinus maenas*. Comp Biochem Physiol 77A: 635–639
Söderhäll K (1981) Fungal cell walls β-1, 3-glucans induce clotting and phenoloxidase attachment to foreign surface of crayfish hemocyte lysate. Dev Comp Immunol 5: 565–573
Söderhäll K (1982) Prophenoloxidase activating system and melanization – a recognition mechanism of arthropods? A review. Dev Comp Immunol 6: 601–611
Söderhäll K, Ajaxon R (1982) Effect of quinones and melanin on mycelial growth of *Aphanomyces astaci*, a parasite on crayfish. J Invertebr Pathol 39: 105–109
Söderhäll K, Cerenius L (1992) Crustacean immunity. Annu Rev Fish Dis 3–23
Söderhäll K, Smith VJ (1983) Separation of the haemocyte populations of *Carcinus maenas* and other marine decapods and prophenoloxidase distribution. Dev Comp Immunol 7: 229–239
Söderhäll K, Smith VJ (1986) The prophenoloxidase activating system as a recognition and defense system in arthropods. In: Gupta AP (ed) Hemocytic and humoral immunity in arthropods. Wiley, New York, pp 251–286
Söderhäll K, Vey A, Ramstedt M (1984) Haemocyte lysate enhancement of fungal spore encapsulation by crayfish haemocytes. Dev Comp Immunol 8: 23–29
Söderhäll K, Levin J, Armstrong PB (1985) The effect of β-1, 3-glucans on blood coagulation and amoebocyte release in the horseshoe crab, *Limulus polyphemus*. Biol Bull 169: 661–674
Söderhäll K, Smith VJ, Johansson MW (1986) Exocytosis and uptake of bacteria by isolated hemocyte populations of two crustaceans: evidence for cell co-operation in the defense reactions of arthropods. Cell Tissue Res 245: 43–49
Söderhäll K, Rögener W, Söderhäll I, Newton RP, Ratcliffe NA (1988) The properties and purification of a *Blaberus craniifer* plasma protein which enhances the activation of hemocyte prophenoloxidase by a β-1, 3-glucan. Insect Biochem 18: 323–330
Söderhäll K, Aspán A, Duvic B (1990) The proPO system and associated proteins; role in cellular communication in arthropods. Res Immunol 141: 896–907
Spycher SE, Arya S, Isenman DE, Painter RH (1987) A functional thioester-containing α_2-macroglobulin homologue isolated from the hemolymph of the american lobster (*Homarus americanus*). J Biol Chem 262: 14606–14611
Sullivan JD, Watson SW (1975) Purification and properties of the clotting enzyme from *Limulus* lysate. Biochem Biophys Res Commun 66: 848–855
Suzuki T, Natori S (1985) Purification and characterization of an inhibitor of the cysteine protease from the hemolymph of *Sarcophaga peregrina* larvae. J Biol Chem 260: 5115–5120
Tanada Y, Watanabe H (1982) Studies on colony specificity in the compound ascidian, *Botryllus primigenus* Oka. I. Initiation of 'non-fusion' reactions with special reference to blood cell infiltration. Dev Comp Immunol 6: 43–52
Thörnqvist P-O, Johansson MW, Söderhäll K (1994) Opsonic activity of cell adhesion proteins and β-1,3-glucan binding proteins from two crustaceans. Dev Comp Immunol 18: 3–12
Tokunaga F, Miyata T, Nakamura T, Morita T, Kuma K, Miyata T, Iwanaga S (1987) Lipopolysaccharide-sensitive serine-protease zymogen (factor C) of horseshoe crab hemocytes. Identification and alignment of proteolytic fragments produced during the activation show that it is a novel type of serine protease. Eur J Biochem 167: 405–416
Tsukamoto T, Ishiguro M, Funatsu M (1986) Isolation of latent phenoloxidase from prepupae of the housefly, *Musca domestica*. Insect Biochem 16: 573–581
Tuková L, Rejnek J, Sima P, Ondrejova R (1986) Lytic activities in coelomic fluid of *Eisenia foetida* and *Lumbricus terrestris*. Dev Comp Immunol 10: 181–189
Unestam T, Söderhäll K (1977) Soluble fragments from fungal cell walls elicit defense reactions in crayfish. Nature (Lond). 267: 45–46

Valembois P, Roch Ph, Lassegues M (1988) Evidence of plasma clotting system in earthworms. J Invertebr Pathol 51: 221–228

Valembois P, Seymour J, Roch Ph (1991) Evidence and cellular location of an oxidative activity in the coelomic fluid of the earthworm *Eisenia fetida andrei*. J Invertebr Pathol 57: 177–183

Wright RK, Cooper EL (1983) Inflmmatory reactions of the Protochordata. Am Zool 23: 205–211

Wright RK, Ermack TH (1982) Cellular defense systems of the Protochordata. In: Cohen N, Sigel MM (eds). The reticuloendothelial system; a comprehensive treatise, vol 3, phylogeny and ontogeny. Plenum Press, New York, pp 283–320

Xylander WER (1992a) Immune defense reactions of Myriapoda – a brief presentation of recent results. 8th International Congress Myriapodology, Innsbruck, Austria. July 1990. Ber Nat Med Ver Innsbruck 10: 101–110

Xylander, WER (1992b) Studies on the phenoloxidase of *Rhapidostreptus virgator* (Silvestri 1907) (Diplopoda, Spirostreptidae). 25th Annu Meet Soc Invertebr Pathol, Heidelberg, Germany, Aug 16–21 (Absrt)

Xylander WER, Bogusch O (1992) Investigations on the phenoloxidase of *Rhapidostreptus virgator* (Arthropoda, Diplopoda). Zool Jahrb Physiol 96: 309–321

Xylander WER, Nevermann L (1990) Antibacterial activity in the hemolymph of myriapods (Arthropoda). J Invertebr Pathol 56: 206–214

Yokosawa H, Sawada H, Abe Y, Numakunai T, Ischii S (1982) Galactose-specific lectin in the hemolymph of the solitary ascidian, *Holocynthia roretzi*. Isolation and characterization. Biochem Biophys Res Commun 107: 451–457

Yokosawa H, Odajima R, Ishii S-I (1985) Trypsin inhibitor in the hemolymph of a solitary ascidian, *Halocynthia roretzi*, purification and characterization. J Biochem 97: 1621–1630

Yokosawa H, Harada K, Igarashi K, Abe Y, Takahashi K, Ishii S-I (1986) Galactose-specific lectin in the hemolymph of the solitary ascidian, *Halocynthia roretzi*. Molecular binding and functional properties. Biochim Biophys Acta 870: 242–247

Chapter 4

A Definition of Cytolytic Responses in Invertebrates*

Ph. Roch[1]

Contents

* The original chapter was to have been co-authored with Calogero Canicatti (University of Lecce – Italy). He died suddenly at the age of 42, on January 8, 1993. This chapter honors his memory as a very close friend and enthusiastic colleague.

[1] Laboratoire DRIM, IFREMER-CNRS-Université de Montpellier 2, Case Courrier 80, 2 Place Eugène Bataillon, 34095 Montpellier cedex 5, France

Advances in Comparative and Environmental Physiology, Vol. 23

1 Introduction

Invertebrates, like vertebrates, constitute a wide heterogeneous group continuously confronted by the necessity to maintain their integrity. Although the survival strategies developed by the two groups are similar, the processes of neutralization and elimination of foreign invaders are not based on the same effectors. A multiplicity of factors that act against microorganisms, foreign cells or abiotic material are part of the humoral repertoire of invertebrate defense systems. Many of these factors are naturally occurring, representing innate mechanisms. However, new types of molecules can be elicited in response to invasion (Boman and Hultmark 1987), suggesting the existence of an invertebrate "acquired" immunity. For example, antibacterial activities, not due to lysozyme, are found in invertebrate fluids. In insects, the phylum in which the studies are the most developed, numerous antibacterial peptides are well known; some of them, such as the defensins and cecropins, have counterparts in both insects and mammals (Eisenhauer et al. 1989; Lee et al. 1989).

Nearly all invertebrate phyla possess molecules that lyse vertebrate red blood cells and consequently they are called **hemolysins**. Among the acelomates, hemolytic activity has been observed in tissue extracts (Kamiya et al. 1985; Bernheimer and Rudy 1986). A similar activity is also present in celomate invertebrates, mainly located in the circulating fluid of annelids, arthropods, mollusks and echinoderms (DuPasquier and Duprat 1968; Weinheimer et al. 1969; Day et al. 1972a; Bretting and Renwrantz 1973; Parrinello et al. 1979; Roch 1979a; Anderson 1980; Cenini 1983; Tuckova et al. 1986; Canicatti 1987a). Lytic activity is generally directed against numerous erythrocytes and certain tumor cells and it varies from one species to another. Annelids seem to possess the most powerful lytic activity as revealed by hemolytic titers that reach 40000 units (Roch et al. 1981a). Although vertebrate erythrocytes are definitely not the natural targets of invertebrate hemolysins, they are an ideal cell model to analyze the membrane events that occur during lysis. We postulate that similar lytic mechanisms are used by invertebrates against a large variety of potentially pathogenic invaders (bacteria, protozoa, parasites, etc.) or perhaps even their own altered, self cells.

The interest devoted to the natural lytic components of celomate invertebrates is based on the idea that they do possess complement-like activity. This offers an explanation for the presence of such a complicated system as early as the evolution of fishes and amphibians. Additional information strengthens this view with the recent presentation of possible relationships between invertebrate hemolysins and vertebrate pore-forming proteins (Canicatti 1990). This implicates a different phylogenetic level of relationships between the two systems (Canicatti and Roch 1993). However, since there is no structural nor substantial coding information, the problem is still open to debate.

2 Invertebrate Lytic Activities

In 1968, the presence of hemolytic activity was demonstrated for the first time in the annelid oligochaete *Eisenia fetida* (DuPasquier and Duprat 1968). Naturally found in the celomic fluid of several earthworm species (Kauschke and Mohrig 1987b), the activity was found to be released by a subpopulation of free celomocytes, the chloragocytes, and present in the mucus surrounding the earthworm (Valembois et al. 1988). The cocoon albumin, in which the embryos develop, also contains lytic activity the characteristics of which are different from those of the celomic fluid but identical to those of epithelial glands (Valembois et al. 1984).

2.1 General Parameters

Invertebrate hemolytic activity presents certain constant characteristics which could represent parameters that define its biological capacity (Table 1). Reactive species usually satisfy all these essential parameters but, for unreactive species, negative results must be carefully evaluated. In some cases, the number of available receptors for hemolysin is not appropriate for demonstrating lysis. For instance, untreated sheep erythrocytes are not susceptible to lysis by internal fluids of both a sea star (Leonard et al. 1990) and a fan worm (Parrinello and

Table 1. Common properties of the hemolytic activity in celomate invertebrates

Mediated by protein(s)
Nonenzymatic
Naturally occurrring
Extracellular or tissue localized
Wide reactivity against different red blood cells
Calcium dependence
Thermal inactivation
Fast time course
Sigmoid dose–response curve

Rindone 1981). However, enzymatic desialization or simple supplementation with calcium, have been shown to convert sheep erythrocytes into susceptible targets. Among the mollusks, only the gastropod class possessed lysins against human erythrocytes (Bretting and Renwrantz 1973). In fact, by modifying the experimental conditions such as ionic concentrations, pH or temperature of incubation, it is possible to reveal lytic activity in other classes of mollusks. It is conceivable that hemolysins could be regulated by inhibitors or by enzyme activations. This attractive idea merits a more full investigation since it could represent a mechanism to explain several open questions such as : *how do the reacting species protect themselves from autolysis?*

2.2 Multifunctionality and Occurrence

In addition to their lytic activity, numerous hemolysins possess opsonizing activity, i.e., they are capable of stimulating phagocytosis (Stein and Cooper 1981). In some cases, the presence of hemolysins has been correlated with rosette formation that represents an initial step in recognition before cellular responses (Toupin and Lamoureux 1976; Mohrig et al. 1984). In annelids, hemolysis,

Table 2. Occurrence and origin of echinoderm hemolytic activity

Species	Activity found in					
Echinoids						
Paracentrotus lividus	CF	CL				
Spherechinus granularis	CF					
Lytechinus variegatus			BE			
Anthocydaris crassispina	CF					
Pseudocentrotus depressus	CF					
Hemicentrotus pulcherrimus	CF					
Strongylocentrotus droebachiensis	CF					
Asteroids						
Asterina gibbosa	CF					
Echinaster sepositus	CF		BE	M		
Marthasterias glacialis	CF	CL		M		
Astropecten auriantacus			BE			
Asteria forbesi	CF					
Asterina pectinifera					GO	
Holothurioids						
Cucumaria planci	CF		BE			
Holothuria polii	CF	CL	BE			WVS
Holothuria tubulosa	CF					
Holothuria impatient	CF					
Ophiurioids						
Ophiocoma echinata			BE			

CF, celomic fluid; CL, celomocyte lysate; BE, body extract; M, mucus; WVS, water vascular system; GO, gonad.

antibacterial and agglutinating activities are shared by the same molecular species and can be separately revealed by modifying the assay (Roch et al. 1987). In arthropods, numerous activities, such as antibacterial, agglutination, protease inhibition and hemolysis are normally undetectable but can be induced by various injections that stimulate.

The most explored sources of invertebrate hemolysis have been the celomic fluid and lysates of various celomic cells (celomocytes or hemocytes), although fluid from the water vascular system of echinoderms, several tissue extracts, egg contents and external mucous secretions have been shown to express lytic activities (Table 2).

2.3 Non-Protein Lysins

Invertebrates also possess lytic activities not mediated by proteins but by saponins which are thermostable glycosides or phosphorylglyceryl esters. Hematophage biting insects (arthropods) and leeches (annelids) possess powerful hemolytic activity that is present in the digestive tract and not involved in defense but obviously essential for their own nourishment. Another cytolytic molecule, lombricin, has been isolated from the oligochaete *Lumbricus terrestris* (Nagasawa et al. 1991). Lombricin is a phosphorylated chain of 6 carbons and 3 amine functions, that inhibit the in vivo growth of spontaneous mammary tumors in mice. Toxins and venoms are also capable of lysis. Restricted in special organs or glands, the best known of these are derived from echinoderms and arthropods and have recently been analyzed in connection with immunity (Canicatti and Roch 1993).

3 Cell-Target Susceptibility

3.1 Erythrocytes and Tumor Cells

The available information on hemolysis reveals that hemolytic activity is generally directed against a wide variety of erythrocytes with enormous differences depending upon the species. For instance, rabbit erythrocytes are lysed by celomic fluid of the oligochaete *Eisenia fetida* yielding titers of about 40 000 whereas human, frog, chicken and rat erythrocytes were lysed at titers of 3 000 (DuPasquier 1971). Among the polychaetes, *Glycera dibranchiata* possesses hemolytic activity directed toward both vertebrate and invertebrate cells, except those from annelids (Table 3). Hemolytic titers are lower than in *Eisenia*: 250 for *Petaloproctus terricola* and 128 for *Spirographis spallanzanii* (Roch et al. 1990).

The best-known hemolytic system of arthropods, characterized by a high degree of specificity toward sheep erythrocytes, has been described in the lobster *Palinurus argus*. Insect hemolymph from *Galleria mellonella* is hemolytic toward numerous vertebrate erythrocytes, revealing significant differences between the

Table 3. Cytolytic activity of celomic fluid from the polychaete *Glycera dibranchiata*. (Chain and Anderson 1983)

	Target cells	Lethality
Bacteria	*Serratia marcescens*	+
Annelids	*Nereis* sp.	–
	Lumbricus sp.	–
Mollusks	*Mercenaria mercenaria*	+
Crustaceans	*Gallinectes sapidus*	+
Insects	*Galleria mellonella*	+
	Manduca sexta	+
Mammals	Sheep erythrocytes	+
	Rabbit erthrocytes	+
	Human erythrocytes	+

erythrocytes (Phipps et al. 1989). Generally, the most sensitive erythrocytes are those of rabbits whereas those from sheep are more resistant.

The fluid of the mollusk *Mytilus edulis* is hemolytic against human, swine, rat, mouse, guinea pig and carp erythrocytes, at titers that never exceed 5 (Leippe and Renwrantz 1988) but erythrocytes from cattle, sheep, bullfrogs and eel were not lysed. A potent hemolytic activity has been reported in the gastropods *Aplysia kurodai* (Yamazaki et al. 1989a) and *Dolabella auricularia* (Yamazaki et al. 1989b). Normal cells as well as tumor cells are lysed, but tumor cells are more sensitive. Concentrations as low as 6 to 20 ng/ml of partially purified molecules are enough to induce 100% lysis of tumor cells whereas normal cells required concentrations of greater than 2 000.

In the echinoderm, *Strongilocentrotus droebachiensis*, only rabbit erythrocytes were lysed (Bertheussen 1983). Echinoderm celomic fluid is also lytic toward various tumor cells, e.g., mouse mastocytoma, human leukemia or mouse fibrosarcoma.

3.2 Autologous Cells

In contrast to erythrocytes, autologous cells from annelids have not been found to be sensitive to lysis (Roch et al. 1981a). Similarly, in echinoderms, autologous cells were not killed by lytic fluids as demonstrated by fertilization and segmentation of *Paracentrotus lividus* eggs that had previously been incubated with hemolytic fluid (Canicatti 1991). However, cleavage patterns were not normal resulting in some irregular blastomeres.

3.3 Cell Surface Interaction

The differences in susceptibility most probably depend upon structural characteristics of cell targets that bear no relationships to phylogenetic position. They

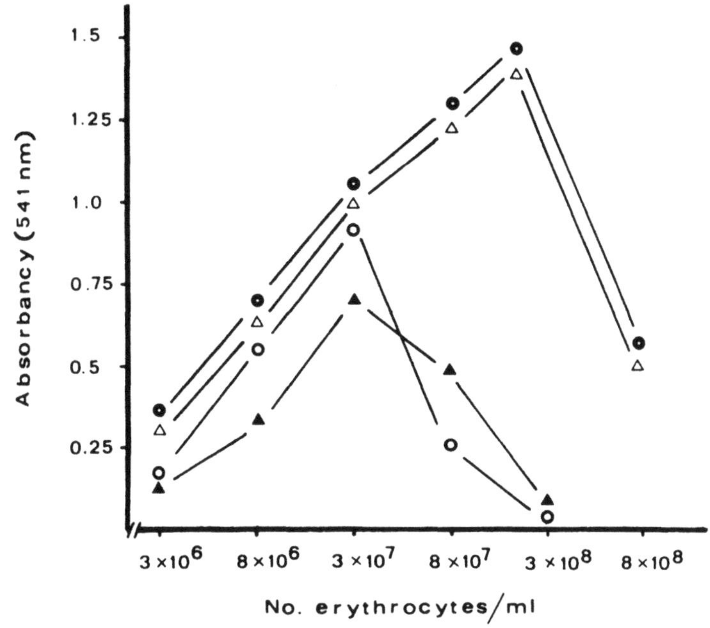

Fig. 1. Hemolytic activity of celomic fluid from the echinoderm *Holothuria polii* as a function of erythrocyte concentration. Erythrocytes were from horse (▲), calf (○), rabbit (Δ) and human group A (●). (Parrinello et al. 1979)

also depend upon the experimental, physicochemical conditions. Meanwhile, using *Holothuria polii* celomic fluid under standard conditions, frog, chicken, pig, horse, calf and sheep erythrocytes underwent hemolysis only when present in diluted suspensions. For each donor species, there is an optimal erythrocyte concentration, independent of the calcium content (Fig. 1). These differences have been interpreted as species-specific differences between putative hemolysin receptors or between hemolysin-receptor interactions (Parrinello et al. 1979). In many species, nucleated erythrocytes from both vertebrates and invertebrates (sipunculid worms for instance) are more resistant than nonnucleated ones (Ryoyama 1973; Parrinello et al. 1979). Nucleated cells could probably escape hemolysis by a number of mechanisms including lipid repair, compensatory ion pumps and protective, membrane-integrated factors.

Of interest is the finding that fungal (zymosan), bacterial (lipopolysaccharides) and algal (laminarin) surface components bind to hemolysins of the echinoderm *Paracentrotus lividus*. As evaluated by residual hemolytic activity, this binding capacity depends upon the concentration of activated particles and could be used to determine the sensitivity of targets (Canicatti 1991). Testing for bactericidal activity of *Asteria forbesi* celomic fluid, Leonard et al. (1990) reported that highly lytic fluids were also bactericidal against different bacteria including the pathogenic *Vibrio tubiashii*. However, in the case of bacterial killing, an important role might be played by other celomic fluid components, such as lysozyme so common in invertebrates (Jollès and Jollès 1975; Lassalle et al. 1988; Canicatti and Roch 1989; Gerardi et al. 1990).

4 Species-to-Species and Individual Variability

In some instances, hemolytic activity has not been revealed even when present in closely related species. This is particularly true in polychaetes [Order Capitellidae (Roch et al. 1990)]. Among the oligochaetes (Order Lumbricidae), *Eisenia fetida andrei* also possesses a powerful hemolytic activity, but most *Lumbricus terrestris, L. castaneus, Allolobophora caliginosa, A. chlorotica, Dendroboena rubida* (Kauschke and Mohrig 1987a) and *Eisenia hortensis,* failed to lyse erythrocytes or displayed light activity. Among the arthropods, only a few species show hemolytic activity and usually with low titers (Table 4). Similarly, hemolysis is not expressed in many mollusks and urochordates, whereas almost all echinoderms possess weak hemolytic activity. Both aquatic and terrestrial invertebrates show hemolytic activity, suggesting that this characteristic is not linked to physiological adaptation. On the other hand, in some groups living in environments with similar or identical pressures, as among the terrestrial oligochaetes or the marine polychaetes, expression of hemolysis is different.

Individual variabilities also exist. For instance, in the annelid oligochaetes *L. terrestris* and *E. hortensis*, not all earthworms show hemolytic activity. The same phenomenon has been reported concerning the mussel *Mytilus edulis* (Leippe and Renwrantz 1988). In echinoderms, the hemolytic activity of echinoids, asteroids and holothurioids also showed individual variations. When the celomic fluids of 78 individuals of the asteroid *Marthasterias glacialis* were tested, only 38 reacted against rabbit erythrocytes (Canicatti 1989). The presence of this activity was

Table 4. Hemolytic activity of celomic fluids and celomocytes in arthropod species

Species	Hemolytic activity
Crustaceans	
Palinurus leniusculus	–
Palinurus argus	+
Palinurus vulgaris	–
Astacus astacus	+
Carcinus maenas	–
Pagurus arrosos	–
Palimnus hirtellus	–
Homarus americanus	+
Insects	
Galleria mellonella	+
Sarcophaga peregrinata	+
Blaberus craniifer	–

independent of sex, age or total protein content in the fluid. The decrease in numbers of reactive sea stars seemed to match the reproductive cycle. Another asteroid, *Asteria forbesi*, also showed individual variations.

One explanation for this variability may be the fact that hemolytic molecules are not continuously released by celomocytes into the circulation; instead the molecules are released in the course of collection and centrifugation procedures. Any stress caused by bleeding induces some of the celomocytes to degranulate and release their lytic activity. Depending upon the rapidity of technical procedures used to collect samples and on cell stability, hemolytic activity has sometimes been recovered in supernatants. In addition, even when collected simultaneously by the same procedure, only some of the earthworms derived from the same, controlled breeding conditions exhibited strong hemolytic activity. Finally, physiological differences between individual worms, such as susceptibility to infection or history of diseases, can also cause stimulation or inhibition of defense reactions.

5 The Hemolytic Reaction

5.1 Kinetics

The time course of hemolytic reactions is rapid since hemolysis has been observed as soon as the erythrocytes are added. After 5 min, the degree of hemolysis reaches a level of 60%. The degree of hemolysis depends on the protein concentration in the fluid, even if lysis is immediate at all concentrations. Plotting the percentage of hemolysis versus serial dilutions of the celomic fluid results in a sigmoid curve (Fig. 2). This pattern is characteristic for many invertebrates (Ryoyama, 1973;

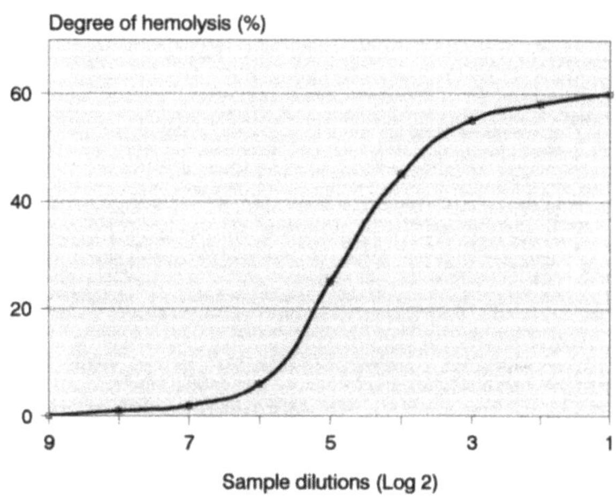

Fig. 2. Sigmoid relationship between percentage of hemolysis and serial dilution of celomic fluid

Roch et al. 1981a; Canicatti and Parrinello 1985; Canicatti 1987a) and resembles the dose–response curve obtained with complement, a complex vertebrate serum component that mediates lysis. Meanwhile, in invertebrates, only one molecular species is responsible for lysis and there is no delay before lysis. In this respect, the invertebrate lytic activity does not resemble the characteristics of complement due to C1 activation. Consequently, the fact that invertebrate hemolysins constitute a simple system, considered together with the resulting sigmoid curve, suggest that hemolysis is mediated by only one molecular species possessing multiple binding sites.

5.2 Stability

Stability through a wide range of pH (4 to 10) and speed of reaction depending upon concentration are properties common to many species. Another common property is the sensitivity to temperature. In the annelid *Eisenia fetida andrei*, hemolysis due to the action of celomic fluid occurs with the same kinetics between 4 and 20 °C, suggesting that the reaction is not due to enzymes (Roch et al. 1981a). Meanwhile, in all annelids tested, lytic activity is sensitive to elevated temperatures and irreversible inhibition occurs at 45–60 °C (Fig. 3). In insects, the hemolytic factor of *Galleria mellonella* has been designated as temperature stable because heating hemolymph at 56 °C for 15 min does not modify this hemolytic capacity (Phipps et al. 1989). In the crustacean *Palinurus argus*, there is no lysis

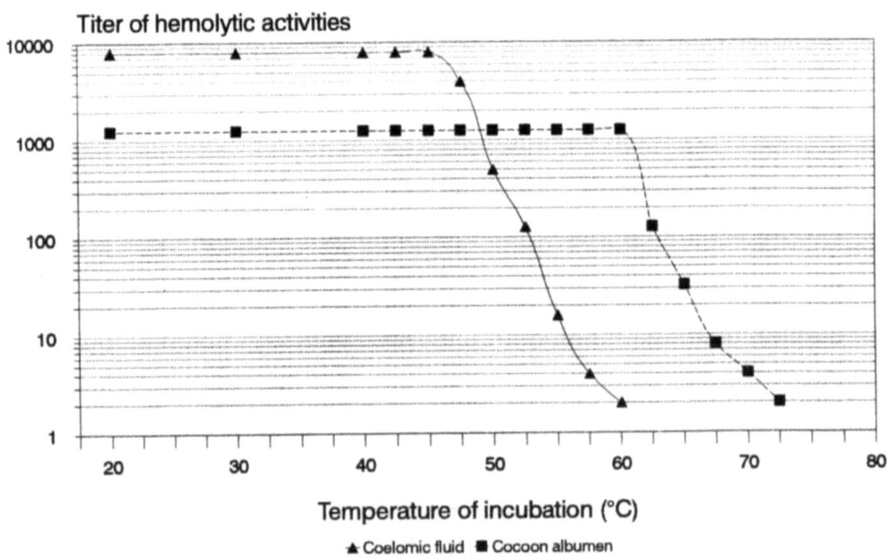

Fig. 3. Temperature stability of hemolysins of the earthworm *Eisenia fetida andrei*. Comparison between hemolytic activity observed in the celomic fluid (*solid line*) and in cocoon albumin (*dashed line*). (Lassègues et al. 1984)

between 0 and 4 °C, revealing the apparent participation of enzymatic processes. Heating the fluid at 52 °C for 20 min results in irreversible loss of hemolytic activity; unfortunately, no biochemical data are available. In echinoderms the hemolytic reaction is also influenced by temperature. Usually, 25–37 °C constitutes the optimal range to obtain the highest degree of hemolysis. For temperatures exceeding 50 °C, hemolytic factors are generally inactivated. The sea star *Marthasterias glacialis* is the one exception in which the hemolytic capacity remains intact even after heating the celomic fluid at 100 °C for 15 min (Canicatti 1989). In the marine mollusk *Aplysia kurodai*, the upper temperature limit is 60 °C (Yamazaki et al. 1989a). As in the annelid *Eisenia fetida andrei* (Roch 1979a), the hemolytic factor of *A. kurodai* is still active after treatment with 8 M urea.

5.3 Divalent Cations

Hemolytic activity in general requires the presence of divalent cations (Ca^{+2}, Mg^{2+}) and is inhibited by chelators such as EDTA. Concerning the lysis of sheep erythrocytes, addition of 10 mM magnesium increases hemolytic titers in the annelids *Nainereis laevigata* and *Orbinia cuvieri* (Roch et al. 1990). For *Petaloproctus terricola*, both calcium and magnesium are necessary. In *Spirographis spallanzanii*, the addition of 20 mM of calcium increases the titer from 2 to 256. This same effect is obtained when calcium is substituted by magnesium (Canicatti and Roch 1993). In *Glycera dibranchiata*, the effect of calcium may be replaced by magnesium. Meanwhile, in some polychaetes, only magnesium can increase lysis. The oligochaete *Eisenia fetida andrei* is an exception since its potent hemolytic activity is not reduced by EDTA (Roch et al. 1981a). Such a fundamental difference could indicate divergence during evolution of hemolytic systems between polychaetes and oligochaetes.

Hemolysins of the insect *G. mellonella* are also independent of divalent cations as the presence of 100 mM EDTA does not modify hemolytic titers. By contrast, in the crustacean, *Palinurus argus*, EDTA totally suppresses hemolytic activity, but adding calcium and magnesium is able to restore such activity, suggesting the participation of divalent cations in the binding of celomic factors onto target cell surfaces. In echinoderms, hemolytic activity has been generally

Table 5. Effect of calcium on percentages of hemolysis of rabbit erythrocytes by echinoderm celomic fluids

	Without calcium	With 10 mM calcium
Echinoids		
Anthocydaris crassispina	2.1	40.7
Strongylocentrotus droebachiensis	0	100
Paracentrotus lividus	0.3	88.5
Holothurioids		
Holothuria polii	0.8	65.5

observed after supplementation with calcium (Table 5). In echinoids (Ryoyama 1973; Bertheussen 1983; Canicatti 1987a), as well as in holothurioids, critical concentrations of calcium, ranging from 10 to 20 mM, shift the degree of hemolysis from 0 to 100%. For instance, in *Holothuria polii*, EDTA as well as heating the amebocyte lysate for 1 h at 45–55 °C, completely abrogates hemolytic activity (Canicatti 1988). At similar concentrations, magnesium cannot substitute the effect of calcium. In contrast, calcium and magnesium are not necessary for the hemolytic activity of *M. glacialis* (Canicatti 1989). Moreover, in the sea star *Asteria forbesi*, dialysis of the celomic fluid against cation-free buffers indicates that neither calcium nor magnesium are required for hemolytic activity (Leonard et al. 1990).

5.4 Anion Regulation

The mechanism by which calcium is involved remains unclear. The presence of calcium could stabilize the structure of lytic molecules or mediate the inter-

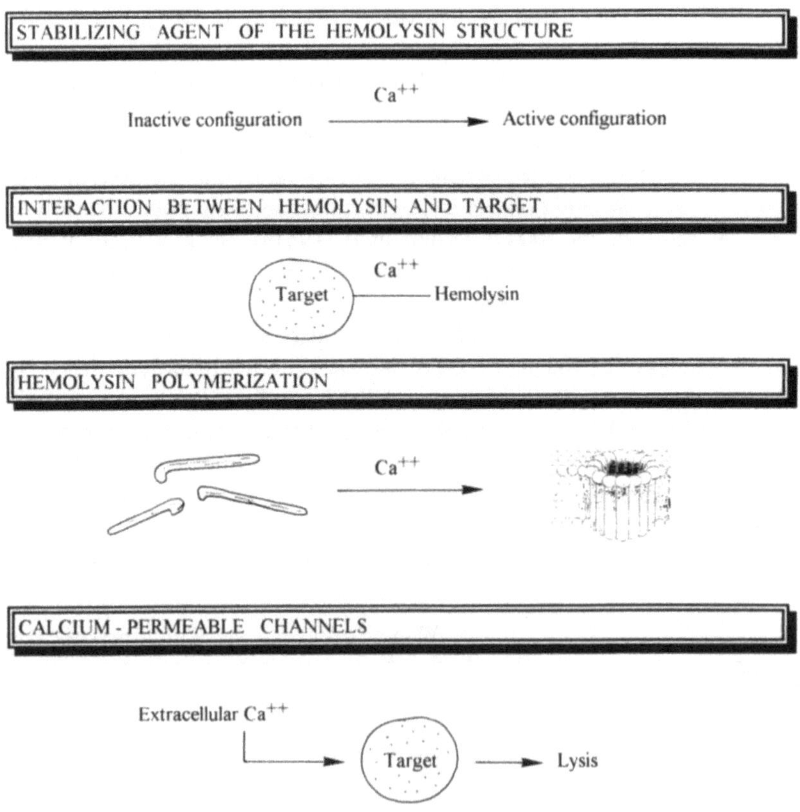

Fig. 4. Possible role played by calcium in erythrocyte lysis mediated by hemolysins

actions between hemolysins and targets (Fig. 4). This would not exclude the involvement of cations in mediating the polymerization of hemolysins causing membrane damage, as happens for the C9 of complement and for perforins (Podack and Tschopp 1984; Podack et al. 1985). On the other hand, as for some actinoporins which are proteic toxins of sea anemones (Michaels 1979), hemolysis could involve the formation of calcium-permeable channels (Ziegler et al. 1986). No other divalent metal cation has been found to be as effective as calcium in increasing the hemolytic titer in echinoderms (Canicatti and Grasso 1988). In fact, in many species, zinc acts as an inhibitor of hemolysis (Leonard et al. 1990; Canicatti 1991; Stabili et al. 1992). At least for *Holothuria polii*, such inhibition is reversible, indicating that the ions are not covalently linked to the hemolytic molecules nor to their putative receptors. It has been suggested that zinc modifies the protein structure by interacting with sulfhydryl groups of hemolysins or with the putative receptors making it impossible for hemolysins to bind efficiently. The antagonistic role of calcium and zinc has recently been considered to be a simple, regulative mechanism (Leonard et al. 1990; Canicatti et al. 1992). If this assumption is correct, the variations in ion composition of internal fluids are critical stimuli for regulating hemolytic activity. Non-physiological conditions, such as stress, wounding, bleeding, pathogen penetration, could easily modify the concentration of ions within the internal fluids. This ultimately will induce the release of active molecules or the modulation of their activity.

6 Membrane Binding and Damage

6.1 Binding Inhibitors

Incubating erythrocyte stroma with hemolymph results in a total loss of lytic capacity, suggesting that hemolysins bind to target cell membranes. Whether or not specific receptors are involved in the binding is unclear. From inhibition studies using carbohydrates, it has been demonstrated that acetylated or methylated saccharides are inhibitors (Bertheussen 1993). In the annelid *Eisenia fetida andrei*, both mannopyranoside and galactosamine prevent lysis when added into the reaction medium. Preincubating the erythrocytes with various carbohydrates revealed that only galactosamine inhibits lysis by interacting at the target cell membrane level (Roch et al. 1981a); (Fig. 5). Lipids are also powerful inhibitors. As observed in annelids (Roch et al. 1989), sphingomyelin is the most powerful inhibitor in sea cucumbers (Canicatti et al. 1987) and sea stars (Canicatti 1989), but not in sea urchins (Canicatti 1987a). Cholesterol and phosphatidyl-ethanol amine are also strong inhibitors in holothurioids (Stabili et al. 1992). The role of lipids is not understood but it is most probably through their interaction that hemolysins are able to penetrate the erythrocyte membrane bilayer, leading to cell disruption.

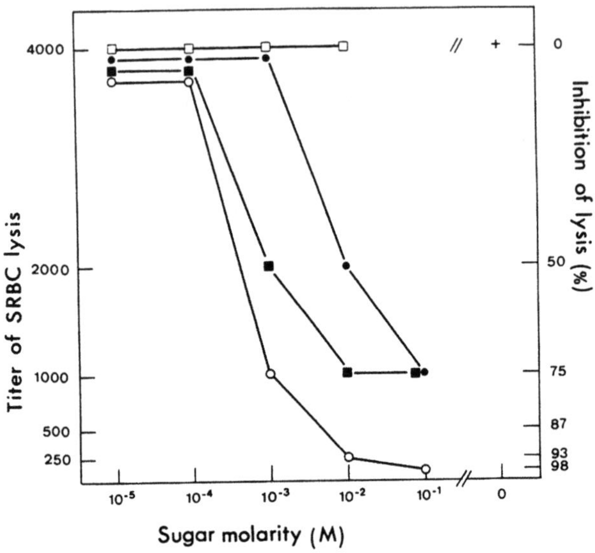

Fig. 5. Effect of various sugars on the lysis of sheep erythrocytes by the celomic fluid of the earthworm *Eisenia fetida andrei*. □: urea, D-arabinose, D-galactose, L-arabinose, lactose, D-fructose; ●: maltose, L-glutamine, D-xylose, L-fucose, β-D-glucose; ■: α-methyl-D-mannopyranoside, N-acetyl-D-galactosamine; ○: N-acetyl-D-glucosamine. (Roch et al. 1981a)

6.2 Transmembrane Channels

The binding is rapid since in less than 2 min, 50% of the hemolysins are already absorbed onto target cell membranes. Using artificial sphingomyelin microvesicles, it has been demonstrated that the hemolysin molecules of annelids undergo a polymerization process that causes the development of high mol. wt. complexes (Roch et al. 1989). Using both echinoderm and annelid hemolysins, erythrocyte membrane disruption results from the formation of transmembrane channels (Canicatti and Tschopp 1989). Lysed rabbit erythrocytes appeared to be covered by irregular holes of heterogeneous sizes ranging from 50 to 250 Å (Canicatti 1987b). Precise observations revealed that the holes are ring-shaped structures that have an average diameter of 10 nm, surrounded by a clear ring (Fig. 6). Such pictures provide strong evidence in favor of a mechanism that involves the formation of transmembrane pores. Transmembrane pore formations are the mode of action of complement (Humphrey and Dourmashkin 1969), perforins (Dourmashkin et al. 1980) and many bacterial toxins (Freer et al. 1968). Meanwhile, to establish a degree of relationship between invertebrate hemolysins and the pore-forming proteins based only on structural homology is highly speculative.

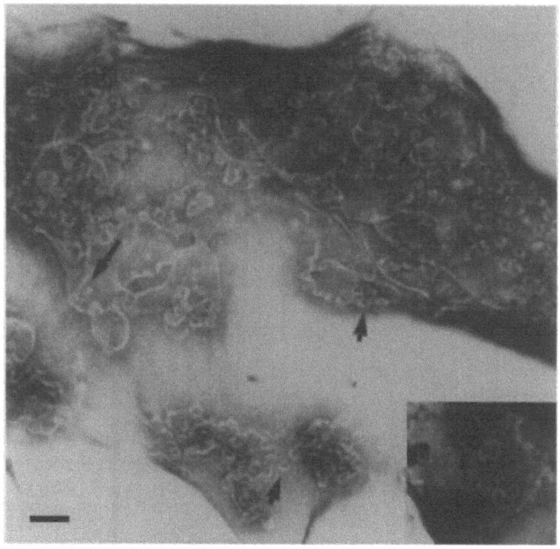

Fig. 6. Membrane damage to sheep erythrocytes by echinoderm *Holothuria polii* hemolysins. Negative staining of lysed membrane. The *arrows* point to circle- and arc-shaped lesions. *Insert* shows a hole of 10 nm in diameter surrounded by a ring of dense material consisting of polymerized hemolysins *Bar* = 50 nm

7 Induction and Regulation

7.1 Stimulation by Injections

Induction phenomena have been investigated extensively in the annelid *Eisenia fetida andrei.* Injections of erythrocytes result in an enormous increase in hemolytic titers (Lassègues et al. 1989a). The induction is maximum at 7 days post-injection and returns to the baseline of activity after 10–12 days (Fig. 7). Even if the activity is always more effective against the erythrocytes used to prime the earthworms, the kinetics are identical to those observed against other living or fixed erythrocytes or bacteria (Lassègues et al. 1989b). Moreover, injections of bovine serum albumin, saline or by a sham injection did not modify the hemolytic titer. Increased hemolytic activity was also reported in earthworms contaminated experimentally by soil pollutants such as polychlorinated biphenyls (PCBs). In fact, immune-related responses mediated by free proteins, such as lysozyme, hemolysins, antibacterial factors and proteases, were simulated (Roch and Cooper 1991). However, those functions that necessitate cell–cell interaction, such as E-rosetting (Rodriguez-Grau et al. 1989), phagocytosis and wound healing (Cooper and Roch 1992), decreased dramatically.

In the polychaete *Neoamphitrite figulus*, a series of erythrocyte injections did not increase but gradually decreased the hemolytic titer (Dales 1982). Such

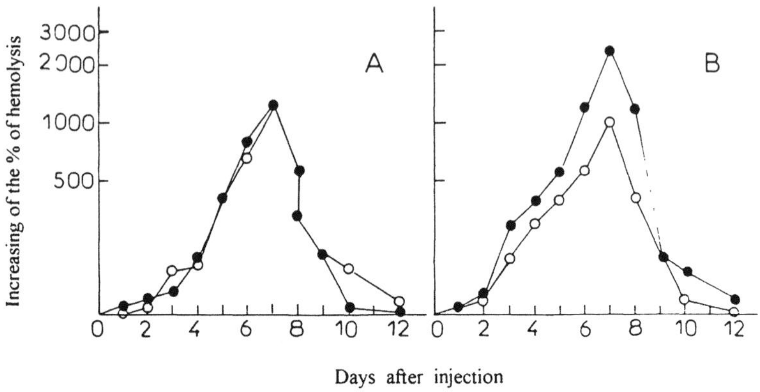

Fig. 7. Kinetics showing induction of hemolytic activity in the earthworm *Eisenia fetida andrei* following one injection of human (**A**) or sheep erythrocytes (**B**). Comparative activity towards sheep (*solid circles*) and human erythrocytes (*open circles*)

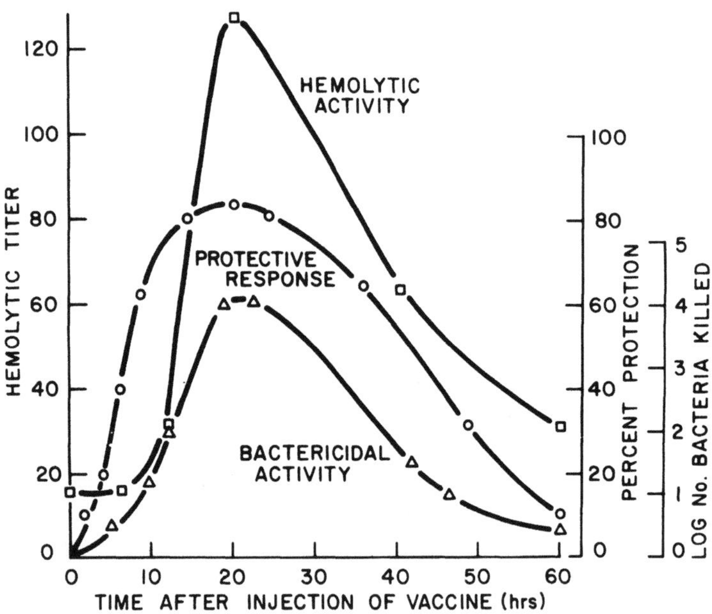

Fig. 8. Parallel between increased hemolytic activity in the insect *Galleria mellonella* and protective immunity and antibacterial activity. (Phipps et al. 1989)

a decrease can result from the absorption of hemolytic molecules onto target cell membranes.

In insects, the hemolytic activity of *Galleria mellonella* is inducible by inoculation of formalin-treated bacteria (Phipps et al. 1989). The in vitro increase in hemolytic activity is related to an increased in vivo protective activity against bacterial infections and to increased in vitro bactericidal activity (Fig. 8). This correlation between bactericidal and hemolytic activities, together with the fact that the hemolysins of oligochaetes are also bactericidal, strengthened the major role of hemolytic molecules in immunodefense. In the larvae of *Sarcophaga peregrina*, hemolymph lectins also act in lysing sheep erythrocytes that have been injected in the body cavity and induction is apparently specific for the type of injected erythrocytes. In mollusks, as in echinoderms, there is no report on induction capability.

7.2 Protein Maturation

Stimulating the annelid *Eisenia fetida andrei* in the presence of actinomycin D, which blocks the transcription of DNA, or cycloheximide, which blocks the translation of RNA, revealed that the transcription-hemolytic genes occurred in 3-4 h (Hirigoyenberry et al. 1990). The translation of the corresponding RNA takes place in less than 24 h. Meanwhile, the maturation of hemolytic proteins requires 3 more days, a long regulatory delay that is not well understood.

7.3 Proteases

Inhibitory experiments involving proteases performed on holothurioids, asteroids and oligochaetes revealed that soybean trypsin inhibitor (STI), benzamidin, phenylmethylsulfonyl fluoride (PMSF) and N-tosyl-L-lysine chloromethyl ketone (TLCK), did not affect hemolysis. Consequently, hemolytic activity may not be due to enzymatic activity at the target cell membrane level. For instance, hemolysins of *Holothuria polii* are not sphingomyelinases (Canicatti et al. 1987). Both the celomic fluid and the celomocytes contain serine-proteases. In *Holothuria polii*, three serine-proteases have been identified by affinity labeling with ^{3}H-DFP. One, the holozyme A, is a 29 kDa protein of a trypsin type (Canicatti and Tschopp 1990). In *Eisenia fetida andrei*, among the three serine-proteases isolated from celomocyte lysates, one of 42 kDa is trypsin type, calcium-independent, one of 48 kDa is chymotrypsin type, and one of 46 kDa is neither the trypsin nor the chymotrypsin type (Roch et al. 1991).

The involvement of serine proteases in hemolytic mechanisms remains unclear. They do not act directly on hemolysis since nonhemolytic fluids remain proteolytic. In *Asterias forbesi*, however, a serine protease could be a component of the asteroid hemolytic system as reported by Leonard et al. (1990). In *Eisenia fetida andrei*, inhibition of the serine proteases resulted in less hemolytic activity, whereas in vitro treatment of cell lysates with serine proteases increased this activity. Those enzymes isolated from earthworms and sea stars are per se nonhemolytic. In mammals, serine proteases are integral components of the complement system (Reid and Porter 1981), the coagulation response and perforin-mediated cytolysis (Tschopp and Nabholz 1990). In invertebrates, serine proteases most probably regulate immune functions, as they activate defense processes in insects (Ashida and Yoshida 1988; Brehélin et al. 1991), crustaceans

(Söderhäll and Smith 1986), echinoderms (Canicatti and Tschopp 1990) and annelids (Roch and Stabili 1992).

8 Hemolytic Molecules

8.1 Annelids

In the polychaetes *Nainereis laevigata* and *Petaloproctus terricola*, hemolysis is due to a single protein which is different in both cases and from those of *Eisenia fetida andrei* (Roch et al. 1990). Only the hemolysin of *Nainereis laevigata* has been isolated by chromatography (Roch and Stabili 1992). The active protein is a 280 kDa tetramer that consists of four identical subunits of 70 kDa (Fig. 9). The mucus of *Spirographis spallanzani* also shows hemolytic activity mediated by at least three major (16, 22 and 40 kDa) and two minor (29 and 34 kDa) proteins that interact with erythrocyte membranes (Canicatti et al. 1992).

In *Eisenia fetida andrei*, the lysis of erythrocytes is mediated by two monomeric lipoproteins of 40 and 45 kDa (Roch et al. 1981a). Separated by isoelectric

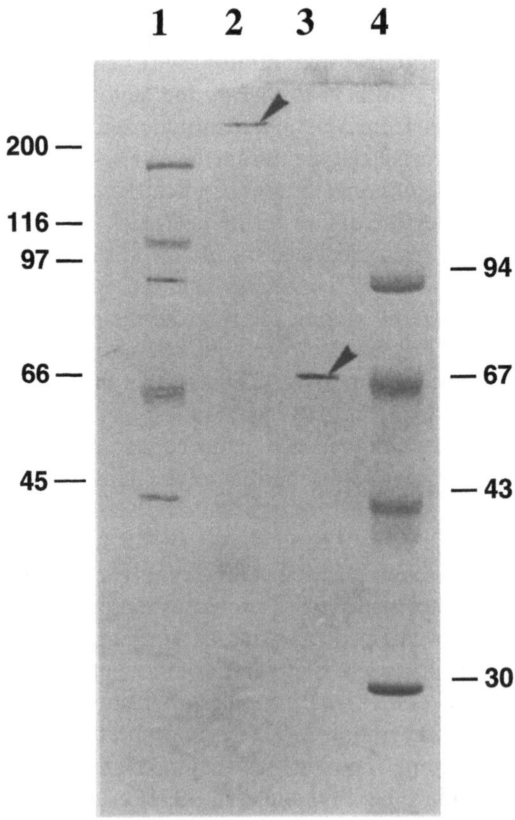

Fig. 9. Tetrameric structure of the hemolysin from the polychaete *Nainereis laevigata*. SDS-PAGE (10%) analysis of hemolytic fraction isolated by anion exchange FPLC of total worm extract. Elution by 300 mM NaCl. *1* and *4* Molecular weight standards; *2* nonreducing conditions, note the presence of a 280 kDa protein; *3* reducing conditions, note the absence of the 280-kDa protein and the presence of a 70-kDa protein. (Roch and Stabili 1992)

focusing, they are polymorphic and consist of five proteins with isoelectric points ranging from 5.9 to 6.3. Analyzing thousands of earthworms as well as their descendence patterns revealed the existence of two independent genes, one of which was always expressed and coded for the pI 6.0 hemolysin, and the second of which possessed four alleles that defined ten genetic families (Roch et al. 1987). They were purified by chromatofocusing (Roch and Davant 1982) and their amino acid compositions showed significant similarities characterized by large quantities of aspartic acid, glutamic acid and glycine (Roch et al. 1986). Structural similarity between the two hemolysins was reinforced by the fact that polyclonal as well as monoclonal antibodies raised against purified hemolysins did not discriminate between them. Hemolysin derived from cocoons of *Eisenia fetida andrei* has also been identified as a native 110 kDa protein (Lassègues et al. 1984). The powerful hemolytic activity found in the *Eisenia* subspecies, *Eisenia fetida fetida*, is mediated by a totally different system: four proteins with mol. wts. that range from 45 to 70 kDa, also polymorphic, but extraordinarily complicated since the number of simultaneously expressed isoforms varied from 2 to 12 (Roch et al. 1980).

8.2 Insects

The hemolysin of the insect, *Galleria mellonella*, has been partially purified by chromatographic separation of the hemolymph. The hemolytic fraction contained two major monomeric molecules of 69 and 75 kDa and a minor molecule of 220 kDa that could represent a polymer (Phipps et al. 1989). In recent years, research concerning the immune system of insects has focused on antibacterial peptides. Since only a few reports have dealt with large hemolytic proteins, this may explain the lack of data.

8.3 Mollusks

Among the mollusks, the biochemical data concerns only the opistobranchia (Table 6). *Aplysia kurodai* contains at least three hemolysins: (1) the aplysianin A, a tetramer of 320 kDa, found in the albumen gland; (2) the aplysianin E, a heterotrimer of 250 kDa, isolated from the eggs; (3) the aplysianin P, a monomer of 60 kDa, found in the celomic fluid (Yamazaki et al. 1984; 1989a). *Aplysia juliana* also contains a different cytolytic glycoprotein of 375 kDa isolated from the genital mass and from the eggs. It was composed of noncovalently bonded subunits of 78 kDa (Kamiya et al. 1988). *Dolabella auricularia* also contained cytolytic proteins: dolabellanin A, a tetramer of 250 kDa (Kisugi et al. 1989) and dolabellanin P, a monomer of 60 kDa (Yamazaki et al. 1989b). Comparing the amino acid compositions of all these molecules revealed close relationships between the aplysianins, whereas the dolabellanins appeared to be unique.

Table 6. Diversity of lytic proteins from three mollusks (opisthobranchia)

Species	Lysin	Mol. wt. (kDa)	Subunits
Aplysia kurodai	Aplysianin A	320	4
	Aplysianin E	250	2
	Aplysianin P	60	1
Aplysia juliana		375	4–5
Dolabella auricularia	Dolabellanin A	250	4
	Dolabellanin P	60	1

8.4 Echinoderms

Three proteins of the echinoderm *Holothuria polii* have been found to be associated with erythrocyte membranes; a high mol. wt. complex, a major 27 kDa protein and a minor 31 kDa protein. Under reducing conditions, the high mol. wt. complex splits into two proteins of 80.5 and 64 kDa suggesting a process of polymerization that leads to the presence of rings that have been observed in ultrastructure (Canicatti 1987b). Among the celomic fluid fractions separated by chromatography, hemolytic activity is eluted before bovine serum albumin. By using celomocyte lysates and erythrocyte overlay techniques, two hemolytic bands have been located and referred to as hemolysin 1 and hemolysin 2 (Canicatti and Ciulla 1988).

9 Hemolysin-Producing Cells and Granules

9.1 Annelids

The celomic cavity of annelids is filled with a liquid that acts not only as a hydroskeleton but also contains numerous, freely mobile cells, called celomocytes. They originate from the internal integument of the celomic cavity but the precise localization of a hemopoietic organ is still controversial (Cooper and Stein 1981). There seems to be an exception for the earthworm *Amynthas diffringens*, in which a cell maturation organ could exist in the smooth muscles that protrude into the celomic cavity.

In the polychaete *Arenicola marina*, at least five morphological classes of celomocytes have been described: spherical, spindle-shaped, lymphocyte-like cells, small and large granulocytes (see Dales and Dixon 1981, for review). In most instances, the relationships between different types of polychaete celomocytes are unknown.

In oligochaete Lumbricidae, three celomocyte lineages have been defined (Valembois 1971); (1) macrophages, which represent 55% of the celomocytes, have been shown to participate in inflammatory reactions (Roch 1979b) and wound healing (Cooper and Roch 1984), and to be capable of phagocytosis and degradation of foreign proteins (Valembois et al. 1973; Tuckova et al. 1986; Mohrig et al. 1989; Bilej et al. 1993); (2) the leukocytes, which are 5–10 μm (35%), mediate natural cytotoxicity (Valembois et al. 1980), release a lysozyme (Lassalle et al. 1988), recognize allogeneic antigens, proliferate (Valembois and Roch 1977) and can be stimulated by various mitogens (Roch 1977); (3) chloragocytes, which are the largest cells (50 μm) and have a small nucleus and a voluminous cytoplasm filled with granules called chloragosomes. Free chloragocytes represent 10% of the celomocytes and are old cells whose maturation occurs when they invest the internal integument. They participate in numerous functions: nutrition, excretion and respiration; since they have been shown to synthesize the respiratory pigment known as hemerythrin (Valembois and Cazaux 1970). Elegant experiments using the hemolytic plaque assay have shown that chloragocytes release hemolytic proteins (DuPasquier and Duprat 1968). Using polyclonal antibodies, young chloragocytes, still attached to the intestine, are capable of synthesizing hemolysins that are stored in the chloragosomes. Whenever trauma occurs, the chloragosomes are released into the medium where hemolytic activity can be observed. This release is initiated when harvesting the celomocytes, but it also occurs naturally when small amounts of celomic fluid and celomocytes are continuously extruded through the dorsal pores of the integument (Valembois et al. 1985). Mixed with the mucus that surrounds the epithelium, the cytolytic proteins constitute an external barrier that protects earthworms from invaders some of which may be eventually pathogenic. Such proteins are also antibacterial (Roch et al. 1981b; Cotuk and Dales 1984), and perhaps their accumulation in the soil can modify the terrestrial microflora and, consequently, the soil's composition.

About 10% of the leukocytes express hemolysins on their outer cell membrane. When bound with antibodies, the leukocyte/antibody complexes undergo membrane redistribution which results in patching, capping and internalization (Valembois et al. 1978) (Fig. 10). In contrast to the situation in mammals, this redistribution process is only slightly delayed by a lower temperature than what is considered to be optimal for survival of ectotherms.

The nature and function of cells from hemocelomic vessels of leeches (hirudinea) has received little attention and basically, two cell types have been described. (1) The amebocytes that resemble leukocytes are capable of phagocytosis and participate in the elimination of waste by the nephridia. (2) The chloragocytes, less frequent, contain lipid droplets, pigment inclusions and participate at least in nutrition (Sawyer and Fitzgerald 1981).

9.2 Arthropods

Several attempts have been made to establish a convenient classification of arthropod hemocytes but the results are controversial due to the enormous diversity of cell types. Briefly, the insect hemolymph contains prohematocytes,

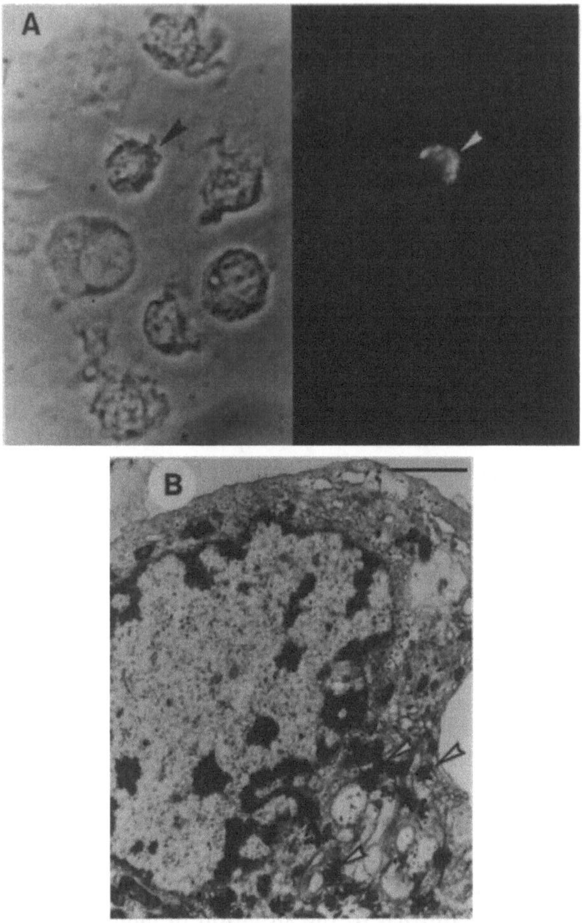

Fig. 10. Membrane redistribution of anti-hemolysin receptors on earthworm leukocyte membranes. **A** Only 10% of the leukocytes express hemolysin (*arrowhead*). Observation in phase contrast and immunofluorescence. **B** After binding, the complexes undergo redistribution followed by internalization (electron microscopy, *bar* = 1 μm). (Canicatti and Roch 1993)

plasmatocytes, granulocytes, spherulocytes, oenocytoids, coagulocytes and adipohemocytes. Embryonic hemocytes are of mesodermic origin and further granulocyte differentiation into other cell-types takes place during postembryonic development (Bahadur 1993). There is no agreement among authors about the types of hemocytes that originate from the hemopoietic organ. Hyaline and semigranular cells of crustaceans, as well as plasmatocytes of insects, are capable of phagocytosis (Bauchau 1981; Rowley and Ratcliffe 1981). The main defense reaction extensively studied in arthropods concerns the encapsulation and melanization processes where granular cells play an active role (see Söderhäll and Smith 1986 for review).

9.3 Mollusks

In the molluskan bivalve *Mytilus edulis*, hemolysin molecules are released by hemocytes and hemolytic activity is detectable in supernatants in short-term cultures of adherent hemocytes (Leippe and Renwrantz 1988). The activity is due to an active secretory process and not to the death of hemocytes. However, the precise events that occur during the release of hemolysins remain speculative.

9.4 Echinoderms

In echinoderms, celomocytes are found in the celomic fluid and in the hemal and water vascular system. They play a number of roles, including immunorecognition and defense against foreign pathogens (Karp and Coffaro 1982). Amebocytes, spherule cells, vibratile cells, hemocytes, crystal cells and progenitor cells are the most common morphological types (Smith 1981). The nature of the hemolysin-producing cells has been established by separating the celomocyte populations using gradient centrifugation. In *Holothuria polii* (Canicatti et al. 1988) as well as in *Paracentrotus lividus* (Pagliara and Canicatti 1992) amebocytes are the sites where hemolysins are stored. As revealed by a rabbit antihemolysin antiserum, both petaloid and filopodial stages of amebocytes show uniform staining of the cytoplasm. Such cells present the general features of phagocytes and constitute the majority of the circulating celomocytes. In *Holothuria polii*, they possess a nucleus, usually spheroidal, with a single nucleolus (Canicatti et al. 1989). The rough endoplasmic reticulum and the Golgi apparatus are well

Table 7. Some characteristics of echinoderm hemolysin-producing cells

Morphology – Large cells with variable shapes – Extremely motile, transitional phases
Cytochemistry – Weakly reactive – Positive for acid and alkaline phosphatases – Positive for chloroacetate-esterase – Positive for phenoloxidase
Function – Phagocytic cells – Cytotoxic in allo- and xenocombinations – Cell aggregation – Encapsulation
Surface phenotypes – Expressing opsonin receptors – Anti-NKH1 positive – Anti-Leu 7 positive – Anti-Leu 19 positive

developed. Moreover, multivesicular bodies, lysosomes and electron-dense material are frequently observed in the vesicles. Vacuoles that contain previously phagocytosed material are also present. Several enzymatic activities have been observed inside the cytoplasm including acid and alkaline phosphatase, chloroacetate-esterase (D'Ancona and Canicatti 1990) and phenoloxidase (Sammarco and Canicatti 1992). The characteristics of hemolysin-producing cells of echinoderms are summarized in Table 7.

Cytoplasmic granules with calcium-dependent lytic activity have been isolated from enriched populations of amebocytes from sea urchins (Pagliara and Canicatti 1992). Amebocytes were disrupted by sonication and then the granules were purified and concentrated on Precoll density gradients (Fig. 11). The fractions that contain hemolytic activity were also shown to possess arylsulfatase activity of type B. This enzyme has been reported to be a possible marker for cytotoxic granules from both vertebrate NK cells (Zucker-Franklin et al. 1983) and cytotoxic T-lymphocytes (Masson and Tschopp 1985). The functional

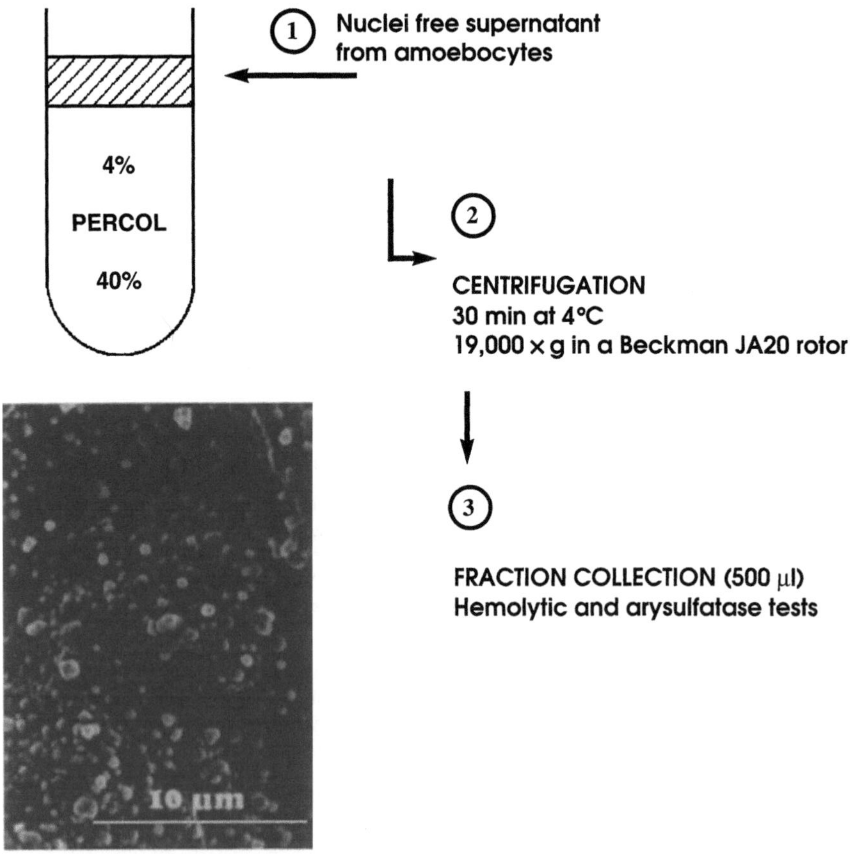

Fig. 11. Experimental procedure used to purify the cytolytic arylsulfatase positive granules in sea urchin and their appearance in scanning electron microscopy

implication of these results suggests that echinoderm amebocytes are not only phagocytic but also cytotoxic. The granules containing the hemolytic activity were homogeneous organelles with variable sizes (0.10–0.25 μm) and were found to be agglutinated in large clusters that never fused.

The mechanism by which amebocytes accomplish cell-mediated lysis is still obscure. The only known mechanism, as in other cellular cytolytic systems, is that a close contact between amebocytes and targets is essential for killing; any later steps remain unknown.

10 Relationship with Cytotoxicity

10.1 Natural Killing and Cytotoxicity

Cytotoxic reactions that are mediated by natural killer cells have been considered as an ancestral immunodefense system and have been observed in several invertebrates (Boiledieu and Valembois 1977; Bertheussen 1979; Söderhäll et al. 1985; Cooper et al. 1992). Invertebrate cytotoxicity is elicited when allo- as well as xenogeneic cells are confronted together in vitro.

In oligochaetes, as well as in sipunculid worms (Weinheimer et al. 1970), circulating leukocytes have been shown to spontaneously lyse erythrocytes, i.e., without having had previous contact. Lysis has been observed by time-lapse microcinematography revealing that several categories of leukocytes could each, separately, effect recognition and lysis (Boiledieu and Valembois 1977) by an unknown mechanism that required close contact between killer and target cells (Kauschke and Mohrig 1987c). The kinetics of lysis was identical regardless of the allo- or xenogeneic combination. Within 6 h, 45% of xenogeneic and 25% of allogeneic target cells were lysed. Maximal reactions remained identical for ratios of 1/25 to 1/50 between target and effector cells. Cytochalasin B, which affects microfilament functions, and EGTA, a chelator of calcium ions, did not inhibit cytolysis. In contrast, cytotoxicity was suppressed by several intracelomic injections of erythrocytes (Fig. 12). Meanwhile, the leukocytes from worms preimmunized with erythrocytes maintained their toxic activity against erythrocytes from other allo- or xenogeneic sources suggesting that natural cytotoxicity was a consequence of specific recognition.

In arthropods, cytotoxic activity has been investigated by using mammalian tumor cells as targets. For instance, leukocytes from the crustacean *Parachaeraps bicarenatus* rapidly lysed Ehrlich cells. Maximum cytotoxicity was reached after 90 min, independently of effector/target cell ratios (Söderhäll et al. 1985). These results suggest that most circulating leukocytes of *P. bicarenatus* are capable of cytotoxicity. The cytotoxic effector factors are different from those of vertebrates but membrane protein determinants seemed to be involved since treatment of crustacean leukocytes with trypsin suppressed cytotoxic reactions. In another crustacean, the crayfish, *Astacus astacus*, at least two hemocyte populations, the granular and the semigranular cells, were cytotoxic (Söderhäll et al. 1985) but

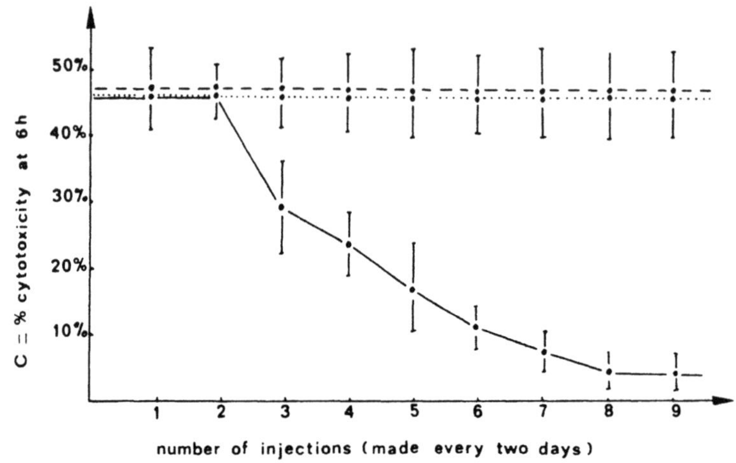

Fig. 12. Decrease in natural cytotoxic capacity of sipunculid marine worms. Effector leukocytes harvested from *Sipunculus nudus* repeatedly injected with target erythrocytes from *Siphonosoma arcassonense* (*solid line*). Natural cytotoxicity remained unchanged toward erythrocytes from another *S. arcassonense* (*dotted line*) or from another sipunculid species, *Phascolosoma vulgare* (*dashed line*), suggesting the existence of specific target recognition. (Boiledieu and Valembois 1977)

incapable of lysing erythrocytes. It must be emphasized that these cytotoxic cells also contain the enzymatic prophenoloxidase system that probably represents the primary recognition and immune effector system of arthropods (Söderhäll 1982).

The amebocytes of echinoderms are phagocytic cells that are also capable of recognition in vitro. At least 70% of the allogeneic cell combinations of sea urchin *Paracentrotus droebachensis*, *P. pallidus* and *Echinus esculentus*, were cytotoxic. These percentages increased to 90 in xenogeneic combinations (Bertheussen 1979). Most of the target cells were lysed after 20 to 35 h of incubation, in an intense two-way reaction, independently of the presence of other cell types, indicating that the amebocytes were the only effector cells. In *Strongilocentrotus granularis*, as well as in *Holothuria polii*, the capacity to damage mammalian tumor cells has also been demonstrated (Stabili et al. 1992). The mouse mastocytoma P 815 was found to be the most sensitive cell type affected by lysis which reached 100% after 60 min of contact with sea urchin amebocytes. A linear relationship was observed between dilutions of celomic fluid and degrees of lysis. In contrast to the lytic activity observed toward sheep erythrocytes, the activity against the human leukemia cells, K 562, was calcium independent. The cytotoxic activity was only slightly reduced by using heated celomic fluid as incubating medium whereas hemolytic activity was dramatically reduced. By immunofluorescence, it has been observed that parasites neutralized inside encapsulating structures in *Holothuria polii*, were covered by hemolysins (Canicatti and Quaglia 1991). This evidence reinforces the view that hemolysins are involved in cytotoxicity. In *Paracentrotus lividus*, auto- and allogeneic cells were neither lysed nor damaged suggesting that the cells either lack hemolysin-binding sites or that they possess certain protective components.

10.2 Lectin Mediation

A variety of mono- and disaccharides blocked the spontaneous cytotoxic reaction, suggesting the involvement of a membrane-integrated lectin as an intermediary essential for target recognition (Decker et al. 1981; Boiledieu and Valembois 1977). For instance, in the sea star *Pisaster gigantus*, 25 mM of arabinogalactane, D-galactose, gentobiose, L-fucose or L-sorbose totally suppressed the cytolysis of horse erythrocytes but not human erythrocytes.

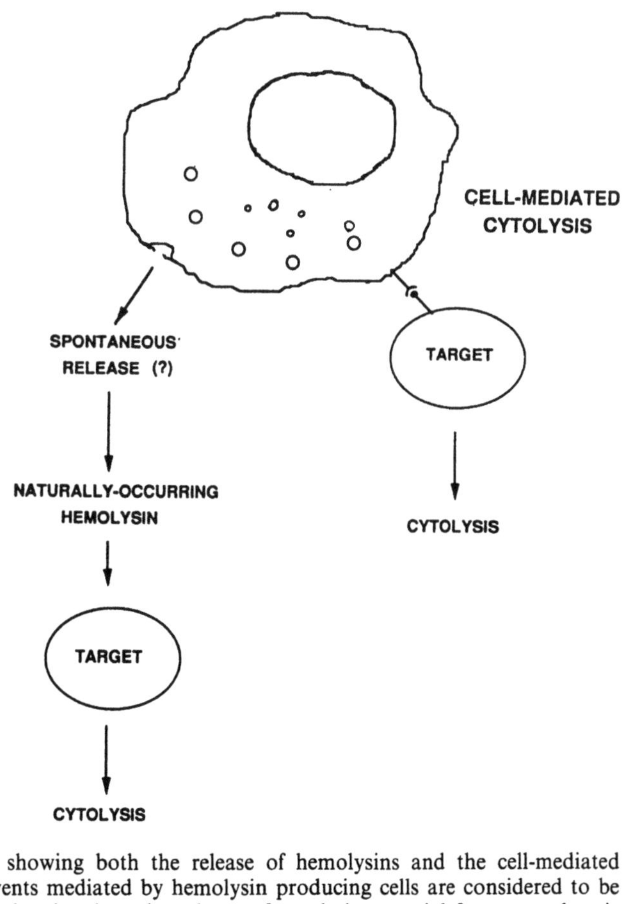

Fig. 13. Proposed model showing both the release of hemolysins and the cell-mediated cytotoxicity. Cytotoxic events mediated by hemolysin producing cells are considered to be a secretory phenomenon that involves the release of cytolytic material from cytoplasmic granules after signaling via a receptor on the effector cell membrane. By a secretory phenomenon, not triggered by target-effector cellular interaction, granules can release lytic molecules into the medium. These molecules, reported as naturally occurring hemolysins, represent one of the humoral components responsible for an immediate response against a variety of invaders

10.3 Effector Analogy

Cytotoxicity mediated by mammalian lymphocytes is basically a secretory phenomenon involving the release of cytotoxic effector components from cytoplasmic granules of effector lymphocytes after signaling via receptors on the effector cell-surface membrane. The finding that hemolysins can be detected in granules of invertebrate phagocytes is consistent with a mechanism associated with exocytosis that is responsible for both secretion of hemolysins and cell-mediated cytotoxicity (Fig. 13). Except for annelids where hemolysins are stored in chloragocytes that are not phagocytic, this implicates certain functional analogies between cytotoxic lymphocytes and invertebrate, hemolysin producing cells. The major reason for not accepting this hypothesis is the lack of any solid evidence for gene coding of invertebrate lytic proteins. Moreover, the capacity of hemolysins to lyse cells does not exclude the possibility that other molecules could share the cytotoxic function.

11 Are Hemolysins Ancestor Immune Molecules?

11.1 Complement Components

Based upon indirect evidence, such as temperature sensitivity, calcium and magnesium requirements and an understanding of various activations or inhibitions, some invertebrate immunologists have recognized in hemolytic activity the presence of what they consider ancestral vertebrate complement (Fig. 14).

In annelids, the proteins C3, C3b, C3-convertase and even C9 have been reported (Bertheussen 1983; Laulan et al. 1983). Meanwhile, zymosan, inulin and LPS, all specific inhibitors of C3b and, consequently, of the alternative complement pathway, are ineffective in hemolysis as has been observed in the earthworm *Eisenia fetida andrei* (Roch et al. 1989). Nucleophils such as hydrazine and methylamine, that inhibit C3 by opening the thioester bond, were also ineffective. The pores produced by earthworm and echinoderm hemolysins and by the C9 showed similar morphologies (Canicatti and Tschopp 1989). Meanwhile, inhibitors of C9 and of perforins did not reduce the hemolytic titers.

The hemolymph of the insect, *Galleria mellonella*, possesses a strong enzymatic activity that can split the chain of bovine C3 just as the C3-convertase does, after it has been activated by cobra venom factor (CVF) (Chadwick et al. 1980; Phipps et al. 1987). This "C3-convertase" of insects was not inhibited by EDTA and it displayed several analogies with those of vertebrates: (1) it was temperature sensitive; (2) it had no effect on the β chain of C3; (3) its activity gave products identical to C3b; (4) it could bind to CVF like mammalian factor B. These significant results argue in favor of an insect hemolymph component similar to factor B and to C3 of the mammalian alternative complement pathway. Unfortunately, the natural substrate of a so-called insect C3-convertase has not yet been identified.

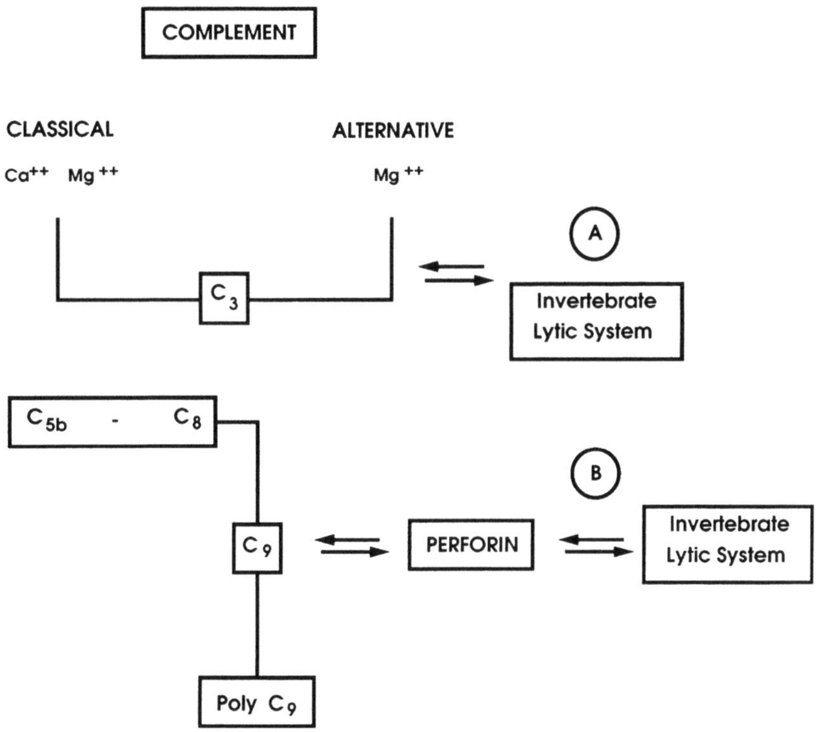

Fig. 14. Speculative phylogenetic relationships between invertebrate (hemolysins) and vertebrate (e.g., complement, perforins) lytic systems. Two major hypotheses were developed during the last twenty years. **A** Existence of a complement-like cascade in invertebrates, activated as the alternative pathway suggesting that invertebrates do possess C3-like proteins. **B** At least functionally, invertebrate hemolysins are related to the perforins of vertebrate NK-cells and cytotoxic T-lymphocytes which, in turn, are related to C9. Moreover, both systems may not be related phylogenetically but evolved as a consequence of convergent evolution. Both would represent independent pathways developed by invertebrates (perhaps including several mechanisms) and vertebrates in response to the commonly shared necessity to preserve life by destroying foreign materials

Molecular weight and functional similarities between the major larval serum protein of the insect *Spodoptera frugiperda* and mammalian C4 binding protein have been reported (D'Cruz and Day 1985). Identical chymotryptic cleavage patterns and the ability of the resulting peptides to prevent assembly of classical C3-convertase were interpreted as an indication that this protein is homologous with the C4bp of the mammalian complement system. Meanwhile, the in vivo immune function of such storage proteins remains unclear, particularly when considering that it represents half the total protein content of hemolymph.

Concerning the mollusks, it has been reported that plasma of the gastropod *Helix pomatia* can activate factor B and C3 (Koch and Nielsen 1984). Here, also the conclusions strongly suggested the presence of complement-proteins in molluscan plasma.

In echinoderms, the starfish *Asterias forbesi* possesses a protein that can cleave purified human C3 and it may be a C3-like proactivator (Day et al. 1972a). Complement-like activity has also been observed in the celomic fluid of the sea urchin *Strongylocentrotus droebachiensis* (Bertheussen 1983). This activity is specific for rabbit erythrocytes and is destroyed by heating the fluid to 37 °C or by reducing the Ca^{2+} concentration. As rabbit erythrocytes are known to activate the alternative pathway, it has been suggested that sea urchins also possess a kind of alternative complement pathway. In fact, the hemolytic activity of invertebrates has never been reported to be restricted to only one target species but is highly dependent upon the number of species tested. Generally, erythrocytes of sheep and mouse are less sensitive than those of rabbit, horse and human B. This does not implicate control mechanisms analogous to those that regulate the alternative pathway but could be due, for instance, to the presence of different sialic acid residues on target cell membranes. In any case, physicochemical and biological properties of *S. droebachiensis* hemolysins still remain uncertain.

The opsonizing effect observed on *S. droebachiensis* macrophages when erythrocytes were coated with C3b was also presented as strong evidence for a complement cascade in echinoderms (Bertheussen and Seljelid 1982). This functional analogy of C3b effects has been interpreted as due to the presence of a C3-receptor on echinoid phagocytes essential to the complement cascade. In mammals, hydrazine and methylamine interact covalently with the thioester bond of C3, leading to its inhibition but, in the sea cucumber *Holothuria polii*, no covalent binding has been demonstrated.

11.2 Perforins

In spite of strong evidence for the existence of complement, invertebrates hemolysins might be related to perforins (Fig. 14). Such an hypothesis is based upon several observations from numerous species. These include the importance of calcium, polymerization of hemolytic proteins into supramolecular complexes and formation of transmembrane channels (Canicatti and Tschopp 1989; Roch et al. 1989). If this is correct, the hemolytic activity, of at least carthworms and echinoderms, does not reflect an ancestral alternative pathway, rather it results from the assembly of pore-forming proteins, a process that is universal. A system of that nature is as efficient as the complement system but far more simple, and easier to imagine when viewed in the natural context of maintaining individual integrity in ancestral animals.

12 Final Comment

Numerous, sometimes contradictory, biological data have been reported concerning hemolytic–cytolytic activity in invertebrates leading to several interpretations. We now have reached the point where biochemical, and especially molecular biological investigations are required in order to know, for instance, where are the ancestors of the complement proteins and, what are the molecular effectors that have been developed by the diverse invertebrate phyla to destroy pathogenic invaders?

References

Anderson RS (1980) Hemolysin and hemagglutinins in the coelomic fluid of a polychaete annelid, *Glycera dibranchiata.* Biol Bull 159: 259–268

Ashida M, Yoshida H (1988) Limited proteolysis of prophenoloxidase during activation by microbial products in insect plasma and effect of phenoloxidase on electrophoretic mobilities of plasma proteins. Insect Biochem 18: 11–19

Bahadur J (1993) Haemocytes and their population. In: Pathak JPN (ed) Insect immunity. Oxford & IBH Pub, New Delhi, pp 14–32

Bauchau AG (1981) Crustaceans. In: Ratcliffe NA, Rowley AF (eds) Invertebrate blood cells, vol 1. Academic Press, New York, pp 385–420

Bernheimer AW, Rudy B (1986) Interactions between membranes and cytolytic peptides. Biochem Biophys Acta 864: 123–127

Bertheussen K (1979) The cytotoxic reaction in allogenic mixtures of echinoid phagocytes. Exp Cell Res 120: 373–381

Bertheussen K (1983) Complement-like activity in sea urchin coelomic fluid. Dev Comp Immunol 7: 21–31

Bertheussen K, Seljelid R (1982) Receptors for complement on echinoid phagocytes. I. The opsonic effect of vertebrate sera on echinoid phagocytosis. Dev Comp Immunol 6: 423–431

Bilej M, Tuckova L, Rejnek J (1993) The fate of protein antigen in earthworms: a study in vitro. Immunol Lett 35: 1–6

Boiledieu D, Valembois P (1977) Natural cytotoxic activity of sipunculid leukocytes on allogenic and xenogenic erythrocytes. Dev Comp Immunol 1: 207–216

Boman HG, Hultmark D (1987) Cell-free immunity in insects. Annu Rev Microbiol 41: 103–126

Brehélin M, Boigregrain RA, Drif L, Coletti-Previero MA (1991) Purification of a protease inhibitor which controls prophenoloxidase activation in hemolymph of *Locusta migratoria* (insecta). Biochem Biophys Res Commun 179: 841–846

Bretting H, Renwrantz L (1973) Untersuchungen von Invertebraten des Mittelmeeres auf ihren Gehalt an hämagglutinierenden Substanzen. Z Immun Forsch 145: 242–249

Canicatti C (1987a) Evolution of the lytic system in echinoderms. I. Naturally occurring hemolytic activity in *Paracentrotus lividus* (Echinoidea) coelomic fluid. Boll Zool 54: 325–329

Canicatti C (1987b) Membrane damage by coelomic fluid from *Holothuria polii* (Echinodermata). Experientia 43: 611–614

Canicatti C (1988) The lytic system of *Holothuria polii* (Echinodermata). A review. Boll Zool 55: 139–144

Canicatti C (1989) Evolution of lytic system in echinoderms. II. Naturally-occurring hemolytic activity in *Marthasterias glacialis* (Asteroidea) coelomic fluid. Comp Biochem Physiol 93A: 587-591

Canicatti C (1990) Hemolysins: pore-forming proteins in invertebrates. Experientia 46: 239–244

Canicatti C (1991) Binding properties of *Paracentrotus lividus* (Echinoidea) hemolysin. Comp Biochem Physiol 98A: 463–468

Canicatti C, Ciulla D (1988) Studies on *Holothuria polii* (Echinodermata) coelomocyte lysate. II. Isolation of coelomocyte hemolysins. Dev Comp Immunol 12: 55–64

Canicatti C, Grasso M (1988) Biodepressive effect of zinc on humoral effector of the *Holothuria polii* immune response. Mar Biol 99: 393–396

Canicatti C, Parrinello N (1985) Hemagglutinin and hemolysin levels in the coelomic fluid from *Holothuria polii* (Echinodermata) following sheep erythrocyte injection. Biol Bull 168: 175–182

Canicatti C, Quaglia A (1991) Ultrastructure of *Holothuria polii* encapsulating body. J Zool Lond 2224: 419–429
Canicatti C, Roch Ph (1989) Studies on *Holothuria polii* (Echinodermata) antibacterial proteins. 1-Evidence and activity of coelomocyte lysozyme. Experientia 45: 756–759
Canicatti C, Roch Ph (1993) Erythrocyte membrane structural features that are critical for the lytic reaction of coelomic fluid hemolysin. Comp Biochem Physiol 105C: 401–407
Canicatti C, Tschopp J (1989) Relationship between vertebrate channel-forming proteins and invertebrate hemolysins. Dev Comp Immunol 13: 377
Canicatti C, Tschopp J (1990) Holozyme A: one of the serine proteases of *Holothuria polii* coelomocytes. Comp Biochem Physiol 96B: 739–742
Canicatti C, Parrinello N, Arizza V (1987) Inhibitory activity of sphingomyelin on hemolytic activity of coelomic fluid of *Holothuria polii* (Echinodermata). Dev Comp Immunol 11: 29–35
Canicatti C, Ciulla D, Farina-Lipari E (1988) The hemolysin-producer coelomocytes in *Holothuria polii*. Dev Comp Immunol 12: 729–736
Canicatti C, Miglietta A, Cooper EL (1989) In vitro release of biologically active molecules during the clotting reaction in *Holothuria polii*. Comp Biochem Physiol 94A: 483–488
Canicatti C, Ville P, Pagliara P, Roch Ph (1992) Hemolysin from mucus of *Spirographis spallanzani* (Polychaeta: Sabellidae). Mar Biol 114: 453–458
Cenini P (1983) Comparative studies on haemagglutinins and haemolysins in an annelid and a primitive crustacean. Dev Comp Immunol 7: 637–640
Chadwick JS, Aston WP, Ricketson JR (1980) Further studies on the effect and role of cobra venom factor on protective immunity in *Galleria mellonella*: activity in response against *Proteus mirabilis*. Dev Comp Immunol 4: 223–232
Chain BM, Anderson RS (1983) A bactericidal and cytotoxic factor in the coelomic fluid of the polychaete, *Glycera dibranchiata*. Dev Comp Immunol 7: 625–628
Cooper EL, Roch Ph (1984) Earthworm leukocyte interactions during early stages of graft rejection J Exp Zool 232: 67–72
Cooper EL, Roch Ph (1992) The capacities of earthworms to heal wounds and to destroy allografts are modified by polychlorinated biphenyls (PCB). J Invertebr Pathol 60: 59–63
Cooper EL, Stein EA (1981) Oligochaetes. In: Ratcliffe NA, Rowley AF (eds) Invertebrate blood cells, vol 1. Academic Press, New York, pp 75–102
Cooper EL, Rinkevick B, Uhlenbruck G, Valembois P (1992) Invertebrate immunity: another viewpoint. Scand J Immunol 35: 247–266
Cotuk A, Dales RP (1984) The effect of the coelomic fluid of the earthworm *Eisenia fetida* Sav. on certain bacteria and the role of the coelomocytes in internal defence. Comp Biochem Physiol 78A: 271–275
Dales RP (1982) Haemagglutinins and haemolysins in the body fluids of *Neoamphitrite figulus* and *Arenicola marina* (Annelida, Polychaeta). Comp Biochem Physiol 73A: 663–667
Dales RP, Dixon LRJ (1981) Polychaetes. In: Ratcliffe NA, Rowley AF (eds) Invertebrate blood cells, vol 1. Academic Press, New York, pp 35–74
D'Ancona G, Canicatti C (1990) The coelomocytes of *Holothuria polii* (Echinodermata). II. Cytochemical staining properties. Bas Appl Histochem 34: 209–218
D'Cruz OJM, Day NK (1985) Structural and functional similarities between the major hemolymph protein of armyworm and cat C4 binding protein of the complement system. Dev Comp Immunol 9: 541–550
Day N, Geiger H, Finstad J, Good RA (1972a) A starfish hemolymph factor which activates vertebrate complement in the presence of cobra venom factor. J Immun 109: 164–167
Day N, Good RA, Muller-Eberhard HJ (1972b) C_3 proactivator in starfish hemolymph. Fed Proc 31: 788
Decker JM, Elmholt A, Muchmore AV (1981) Spontaneous cytotoxicity mediated by invertebrate mononuclear cells towards normal and malignant vertebrate targets: inhibition by defined mono- and disaccharides. Cell Immunol 59: 161–170
Dourmashkin R, Deteix P, Simone CB, Henkart P (1980) Electron microscopic demonstration of lesions on target cell membranes associated with antibody dependent cellular cytotoxicity. Clin Exp Immunol 42: 554–563

DuPasquier L (1971) Etude comparée d'un facteur cytolytique humoral chez une larve d'amphibien et chez un Oligochète. Arch Zool Exp Gén 112: 81–84

DuPasquier L, Duprat P (1968) Aspects humoraux et cellulaires d'une immunité naturelle non spécifique chez l'oligochète *Eisenia foetida* (Lumbricidae). CR Acad Sci Paris 266: 538–546

Eisenhauer PB, Harwig SSL, Szklarek D, Ganz T, Selsted ME, Lehrer RI (1989) Purification and antimicrobial properties of three defensins from rat neutrophils. Infect Immun 57: 2021–2027

Freer JH, Arbuthnott JP, Bernheimer AW (1968) Interaction of staphylococcal alpha toxin with artificial and natural membranes. J Bacteriol 95: 1153–1158

Gerardi P, Lassègues M, Canicatti C (1990) Cellular distribution of sea urchin antibacterial activity. Biol Cell 70: 153–157

Hirigoyenberry F, Lassalle F, Lassègues M (1990) Antibacterial activity of *Eisenia fetida andrei* coelomic fluid: transcription and translation regulation of lysozyme and proteins evidenced after bacterial infestation. Comp Biochem Physiol 95B: 71–75

Humphrey JH, Dourmashkin RR (1969) The lesions in cell membranes caused by complement. Adv Immunol 11: 75–115

Jollés J, Jolleś P (1975) The lysozyme from *Asterias rubens*. Eur J Biochem 54: 19–23

Kamiya H, Muramoto K, Goto R (1985) Cytotoxicity and hemolysis by an aqueous extract of a marine sponge, *Clathria* sp. proc. Symp on Natural toxins, China 1985, pp 24–28

Kamiya H, Muramoto K, Goto R, Yamazaki M (1988) Characterization of the antibacterial and antineoplastic glycoproteins in a sea hare *Aplysia juliana*. Nippon Suisan Gakkaishi 54: 773–777

Karp RD, Coffaro KA (1982) Cellular defence systems of the Echinodermata. In: Cohen N, Sigel MM (eds) The reticuloendothelial system, vol III. Plenum Press, New York, pp 257–282

Kauschke E, Mohrig W (1987a) The occurrence of bacterioagglutinating, hemagglutinating and hemolytical compounds in the coelomic fluid of different species of European earthworms (Annelida, Lumbricidae). Zool Jahrb Physiol 91: 467–477

Kauschke E, Mohrig W (1987b) Comparative analysis of hemolytic and hemagglutinating activities in the coelomic fluid of *Eisenia foetida* and *Lumbricus terrestris* (Annelida, Lumbricidae). Dev Comp Immunol 11: 331–341

Kauschke E, Mohrig W (1987c) Cytotoxic activity in the coelomic fluid of the annelid *Eisenia fetida* Sav. J Comp Physiol 157B: 77–83

Kisugi J, Yamazaki M, Ishii Y, Tansho S, Muramoto K, Kamiya H (1989) Purification of a novel cytolytic protein from albumen gland of the sea hare *Dolabella auricularia*. Chem Pharm Bull 37: 2773–2776

Koch C, Nielsen HE (1984) Activation of vertebrate complement by *Helix pomatia* haemolymph. Dev Comp Immunol 8: 15–22

Lassalle F, Lassègues M, Roch Ph (1988) Protein analysis of earthworm coelomic fluid. IV – Evidence, activity induction and purification of *Eisenia fetida andrei* lysozyme (Annelidae). Comp Biochem Physiol 91 B: 187–192

Lassègues M, Roch Ph, Cadoret MA, Valembois P (1984) Mise en évidence de protéines hémolytiques et hémagglutinantes spécifiques de l'albumen des cocons du Lombricien *Eisenia fetida andrei*. CR Acad Sci Paris 299: 691–696

Lassègues M, Roch Ph, Valembois P (1989a) Antibacterial activity of *Eisenia fetida andrei* coelomic fluid: evidence, induction and animal protection. J Invertebr Pathol 53: 1–6

Lassègues M, Roch Ph, Valembois P (1989b) Antibacterial activity of *Eisenia fetida andrei* coelomic fluid. 2. specificity of the induced activity. J Invertebr Pathol 54: 28–31

Laulan A, Lestage J, Chateaureynaud-Duprat P, Fontaine M (1983) Mise en évidence de substances contenues dans le liquide coelomique de *Lumbricus terrestris*, possédant des fonctions communes avec celles de certains composants du complément humain. Ann Immunol 134: 223–232

Lee JY, Boman A, Chuanxin S, Andersson M, Jörnvall H, Mutt V, Boman HG (1989) Antibacterial peptides from pig intestine: Isolation of a mammalian cecropin. Proc Natl Acad Sci USA 86: 9159–9162

Leippe M, Renwrantz L (1988) Release of cytotoxic and agglutinating molecules by *Mytilus* hemocytes. Dev Comp Immunol 12: 297–308

Leonard LA, Strandberg JD, Winkelstein JA (1990) Complement-like activity in the sea star *Asterias forbesi.* Dev Comp Immunol 14: 19–30
Masson D, Tschopp J (1985) Isolation of a lytic, pore-forming protein (perforin) from cytolytic T lymphocytes. J Biol Chem 260: 9069–9072
Michaels DW (1979) Membrane damage by a toxin from the sea anemone *Stoichactis helianthus.* I: Formation of transmembrane channels in lipid bilayers. Biochim Biophys Acta 555: 67–71
Mohrig W, Kauschke E, Ehlers M (1984) Rosette formation of the coelomocytes of the earthworm *Lumbricus terrestris* with sheep erythrocytes. Dev Comp Immunol 8: 471–476
Mohrig W, Eue I, Kauschke E (1989) Proteolytic activities in the coelomic fluid of earthworms (Annelida, Lumbricidae). Zool Jahrb Physiol 93: 303–317
Nagasawa H, Sawaki K, Fujii Y, Kobayashi M, Segawa T, Suzuki R, Inatomi H (1991) Inhibition by lombricine from earthworm (*Lumbricus terrestris*) of the growth of spontaneous mammary tumours in SHN mice. Anticancer Res 11: 1061–1064
Pagliara P, Canicatti C (1992) Isolation of cytolytic granules from sea urchin amoebocytes. Eur J Cell Biol 60: 179–184
Parrinello N, Rindone D (1981) Studies on the natural hemolytic system of the annelid worm *Spirographis spallanzanii viviani* (Polychaeta). Dev Comp Immunol 5: 33–42
Parrinello N, Rindone D, Canicatti C (1979) Naturally occurring hemoylsins in the coelomic fluid of *Holothuria polii* Delle Chiase (Echinodermata). Dev Comp Immunol 3: 45–54
Phipps D, Menger M, Chadwick JS, Aston WP (1987) A cobra venom factor (CVF)-induced C_3 convertase activity in the hemolymph of *Galleria mellonella.* Dev Comp Immunol 11: 37–46
Phipps D, Chadwick JS, Leeder RG, Aston WP (1989) The hemolytic activity of *Galleria mellonella* hemolymph. Dev Comp Immunol 13: 103–111
Podack ER, Tschopp J (1984) Membrane attack by complement. Mol Immunol 21: 589–603
Podack ER, Young JDE, Cohn ZA (1985) Isolation and biochemical and functional characterization of perforin 1 from cytolytic T cell granules. Proc Natl Acad Sci USA 82: 8629–8638
Reid KB, Porter RA (1981) The proteolytic activation systems of complement. Annu Rev Biochem 50: 433–464
Roch Ph (1977) Réactivité in vitro des leucocytes du Lombricien *Eisenia fetida* Sav. à quelques substances mitogéniques. CR Acad Sci Paris Sér D 284: 705–708
Roch Ph (1979a) Protein analysis of earthworm coelomic fluid. I. Polymorphic system of the natural hemolysins of *Eisenia foetida andrei.* Dev Comp Immunol 3: 599–608
Roch Ph (1979b) Leukocyte DNA-synthesis in grafted lumbricids: an approach to study histocompatibility in invertebrates. Dev Comp Immunol 3: 417–428
Roch Ph, Cooper EL (1991) Cellular but not humoral antibacterial activity of earthworms is inhibited by Aroclor 1254 (PCB). Ecotox Environ Safety 122: 283–290
Roch Ph, Davant N (1982) Rapid procedure for isolation of earthworm bacteriostatic factor using chromatofocusing. J Chromatogr 240: 552–555
Roch Ph, Stabili L (1992) Tetrameric structure of the polychaete *Nainereis laevigata* (Orbinida) cytolytic protein purified by FPLC. Anim Biol 1: 139–144
Roch Ph, Valembois P, Lassègues M (1980) Biochemical particularities of the antibacterial factor of two subspecies *Eisenia fetida fetida* and *Eisenia fetida andrei.* Am Zool 20: 794
Roch Ph, Valembois P, Davant N, Lassègues M (1981a) Protein analysis of earthworm coelomic fluid. II. Isolation and biochemical characterization of the *Eisenia fetida andrei* factor (EFAF). Comp Biochem Physiol 69B: 829–836
Roch Ph, Valembois P, Lassègues M, Cassand P (1981b) Protective activity of the coelomic fluid of a lumbricid *Eisenia fetida.* Dev Comp Immunol (Suppl) 1/5: 81–86
Roch Ph, Valembois P, Vaillier J (1986) Amino-acid compositions and relationships of five earthworm defense proteins. Comp Biochem Physiol 85B: 747–751
Roch Ph, Valembois P, Lassègues M (1987) Genetic and biochemical polymorphism of earthworm humoral defense. In: Cooper EL, Langlet C, Bierne J (eds) Progress in clinical and biological research, vol 233. Developmental and comparative immunology. Alan R Liss, New York, pp 91–102
Roch Ph, Canicatti C, Valembois P (1989) Interaction between the hemolytic system of the

earthworm *Eisenia fetida andrei* and the SRBC membrane. Biochim Biophys Acta 983: 193–198

Roch Ph, Giangrande A, Canicatti C (1990) Comparison of hemolytic activity in eight species of polychaetes. Mar Biol 107: 199–203

Roch Ph, Stabili L, Pagliara P (1991) Purification of three serine proteases from the coelomic cells of earthworms. Comp Biochem Physiol 98B: 597–602

Rodriguez-Grau G, Venables BJ, Fitzpatrick LC, Goven AJ, Cooper EL (1989) Suppression of secretory rosette formation by PCBs in *Lumbricus terrestris*: an earthworm immunoassay for humoral immunotoxicity of xenobiotics. Environ Toxicol Chem 8: 1201–1207

Rowley AF, Ratcliffe NA (1981) Insects. In: Ratcliffe NA, Rowley AF (eds) Invertebrate blood cells, vol 1. Academic Press, New York, pp 420–488

Ryoyama K (1973) Studies on the biological properties of coelomic fluid of sea urchin. I. Naturally occurring hemolysin in sea urchin. Biochim Biophys Acta 320: 157–165

Sammarco S, Canicatti C (1992) Phenoloxidase and lipidic pigments in enriched *Holothuria polii* coelomocyte populations. In: Scalera-Liaci L, Canicatti C (eds) Echinoderm research 1991. Balkema, Rotterdam, pp 217–220

Sawyer RT, Fitzgerald SW (1981) Hirudineans. In: Ratcliffe NA, Rowley AF (eds) Invertebrate blood cells, vol 1. Academic Press, New York, pp 141–159

Smith VJ (1981) The echinoderms. In: Ratcliffe NA, Rowley AF (eds) Invertebrate blood cells, vol 1. Academic Press, New York, pp 513–562

Söderhäll K (1982) Prophenoloxydase activating system and melanization. A recognition mechanism of arthropods? Dev Comp Immunol 6: 601–611

Söderhäll K, Smith VJ (1986) The prophenoloxydase activating system: the biochemistry of its activation and role in arthropod cellular immunity, with special reference to crustaceans. In: Brehélin M (ed) Immunity in invertebrates. Springer, Berlin Heidelberg New York, pp 208–223

Söderhäll K, Wingren A, Johansson MT, Bertheussen K (1985) The cytotoxic reaction of hemocytes from the freshwater crayfish, *Astacus astacus*. Cell Immunol 94: 326–332

Stabili L, Pagliara P, Metrangolo M, Canicatti C (1992) Comparative aspect of echinoidae cytolysins: the cytolytic activity of *Spherechinus granularis* coelomic fluid. Comp Biochem Physiol 101A: 553–556

Stein EA, Cooper EL (1981) The role of opsonins in phagocytosis by coelomocytes of the earthworm *Lumbricus terrestris*. Dev Comp Immunol 5: 415–425

Toupin J, Lamoureux G (1976) Coelomocytes of earthworms: the T-cell-like rosette. Cell Immunol 26: 127–132

Tschopp J, Nabholz M (1990) Perforin-mediated target cell lysis by cytolytic T lymphocytes. Annu Rev Immunol 8: 279–288

Tuckova L, Rejnek J, Sima P, Ondiejova R (1986) Lytic activities in coelomic fluid of *Eisenia foetida and Lumbricus terrestris*. Dev Comp Immunol 10: 181–189

Valembois P (1971) Etude ultrastructurale des coelomocytes du lombricien *Eisenia fetida* Sav. Bull Soc Zool Fr 96: 59–72

Valembois P, Cazaux M (1970) Etude autoradiographique du rôle trophique des cellules chloragogènes des vers de terre. CR Soc Biol 164: 1015–1018

Valembois P, Roch Ph (1977) Identification par autoradiographie des leucocytes stimulés à la suite de plaies ou de greffes chez un ver de terre. Biol Cell 28: 81–82

Valembois P, Roch Ph, Du Pasquier L (1973) Dégradation in vitro de protéine étrangère par les macrophages du Lombricien *Eisenia fetida* Sav. CR Acad Sci Paris Sér D 277: 57–60

Valembois P, Boiledieu D, Roch Ph (1978) Modifications membranaires induites par des antigènes d'histocompatibilité et par des mitogènes chez certains leucocytes d'Invertébrés. Bull Soc Zool Fr 103: 301–308

Valembois P, Roch Ph, Boiledieu D (1980) Natural and induced cytotoxicities in sipunculidea and annelida. In: Manning M (ed) Phylogeny of immunological memory. Elsevier Amsterdam, pp 47–55

Valembois P, Roch Ph, Lassègues M (1984) Simultaneous existence of hemolysins and hemagglutinins in the coelomic fluid and in the cocoon albumen of the earthworm *Eisenia fetida andrei*. Comp Biochem Physiol 78A: 141–145

Valembois P, Lassègues M, Roch Ph, Vaillier J (1985) Scanning electron microscopic study of the involvement of coelomic cells in earthworm antibacterial defence. Cell Tissue Res 240: 479–484

Valembois P, Roch Ph, Lassègues M (1988) Evidence of plasma clotting system in earthworms. J Invertebr Pathol 51: 221–228

Weinheimer PF, Evans EF, Stround RM, Acton RT, Painter B (1969) Comparative immunology: natural hemolytic system of the spiny lobster *Palinurus argus*. Proc Soc Exp Biol Med 130: 322–326

Weinheimer PF, Acton RT, Cushing JE, Evans EE (1970) Reactions of sipunculid fluid with erythrocytes. Life Sci 9: 145–151

Yamazaki M, Kisugi J, Jkenami M, Kamiya H, Mizuno D (1984) Cytolytic factor in eggs of the sea hare *Aplysia kurodai*. Gann 75: 269–274

Yamazaki M, Kimura K, Kisugi J, Muramoto K, Kamiya H (1989a) Isolation and characterization of a novel cytolytic factor in purple fluid of the sea hare *Aplysia kurodai*. Cancer Res 49: 3834–3838

Yamazaki M, Tansho S, Kisugi J, Muramoto K, Kamiya H (1989b) Purification and characterization of a cytolytic protein from purple fluid of the sea hare *Dolabella auricularia*. Chem Pharm Bull 37: 2179–2182

Ziegler W, Pavlovkin J, Remis D, Pokorny J (1986) The anion/cation selectivity of the syringotoxin channel. Biologia (Bratislava) 41: 1091–1100

Zucker-Franklin D, Grusky G, Yang JS (1983) Arylsulfatase in natural killer cells: its possible role in cytotoxicity. Proc Natl Acad Sci USA 80: 6977–6981

Chapter 5

The Immunoglobulin Superfamily: Where Do Invertebrates Fit In?

Y. Kurosawa and *K. Hashimoto*

Contents

1 Introduction

The accumulation of the amino acid sequences of many proteins and the nucleotide sequences of many genes has allowed us to classify the respective proteins and genes into families, primarily on the basis of homology between sequences, secondarily on the basis of three-dimensional structure, and thirdly on the basis of function. The concept of a gene family is based on the hypothesis that all the genes that exist in contemporary organisms are descendants of sets of putative primordial genes that were present in ancestral organisms, and that all of these genes have been created by multiplication and/or exon shuffling of primordial genes, with subsequent diversification by means of mutations and selection during evolution. According to this hypothesis, genes that encode proteins that belong to the same family are considered to be derived from the same primordial gene, at least with respect to those parts of their genes that correspond

Institute for Comprehensive Medical Science, Fujita Health University, Toyoake, Aichi 470-11, Japan to the relevant portions of the proteins (Hill et al. 1966; Ohno 1970; Green et al. 1993).

The immunoglobulin (Ig) superfamily, originally proposed by Williams (1982), is one of the largest families that have been defined, most of which are in the animal kingdom. A gene family has been defined as a set of genes that encode proteins with structural and functional similarities in restricted biological phenomena, for example, the epidermal growth factor (EGF) family. A superfamily, by contrast, is considered to contain multiple gene families that show structural similarities to one another. The most important characteristic of members of the Ig superfamily is a domain that is referred to as the Ig fold (Edmundson et al. 1975; Amzel and Poljak 1979). One Ig domain consists of 90 to 110 amino acid residues which form a bilayer of antiparallel β-sheets that is stabilized by a conserved S–S bond. Proteins that contain at least one Ig domain can be classified as members of the Ig superfamily. Ig-related sequences were identified initially in the molecules that are expressed on lymphoid cells, such as Ig, the major histocompatibility antigen complex (MHC) and several surface antigens of B and T lymphocytes (Williams 1984). Now, however, proteins with such domains have been identified in many non-lymphoid tissues, including not only those of vertebrate origin but also those of invertebrate origin (Williams 1987; Williams and Barclay 1988, 1989). It is possible to argue that the Ig superfamily is a set of structures involved in cell-surface recognition events and in protein–protein recognition events.

In this review, we will first describe the molecular architecture of the Ig fold. Then, we will discuss evolutionary relationships between Ig, the T-cell receptor (TCR) and the MHC. Then, since, in the case of invertebrates, many of the proteins in the Ig superfamily have been identified in nervous systems (Williams and Barclay 1988, 1989) and since various similarities between the immune system and the nervous system have been observed in addition to the presence of Ig-related molecules in both systems (Bazan 1991; Steinman 1993), we will discuss the relationship between these two systems. Finally, we will focus on the origins of immune systems and the process of their diversification during evolution.

2 Molecular Architecture of the Ig Fold

Determinations of three-dimensional structures of antibodies, as well as comparisons of amino acid sequences among antibodies, β2-microglobulin and Thy-1, led Williams to propose the concept of the Ig superfamily in 1982 (Williams 1982; Williams and Gagnon 1982). Since then many proteins, not only of lymphoid origin but also of non-lymphoid origin , such as neural cell adhesion molecules (N-CAMs) have been shown to belong to the Ig superfamily (Cunningham et al. 1987). Based on the alignment of the primary sequences of amino acids of these proteins, Williams proposed that they can be divided into three categories: the V-set, the Cl-set and the C2-set, as discussed in further detail below (Williams 1987; Williams and Barclay 1988, 1989).

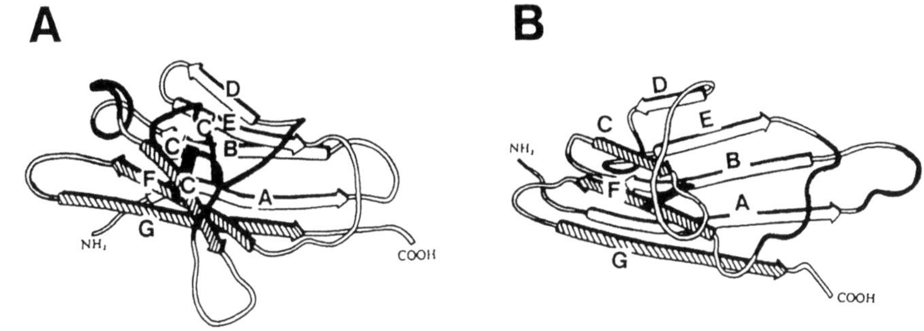

Fig. 1A–B. Structure of the Ig fold. **A** The V domain. **B** The C domain. An S–S bond is indicated by a *thick bar*. (Edmundson et al. 1975; Williams and Barclay 1989)

Three-dimensional structural analysis of antibodies indicated that the structure of variable (V) domains and that of constant (C) domains of antibodies are basically similar to but slightly different from each other (Alzari et al. 1988). As shown schematically in Fig. 1, both domains consist of bilayers of antiparallel β-sheets stabilized by an S–S bond between two conserved cysteine residues. The numbers of β strands that form each sheet differ slightly between V domains and C domains. In the case of C domains, the first sheet is composed of four β strands, A, B, D and E, and the second is composed of three strands, C, F and G. In the case of V domains, the distance between C and D strands is longer than that in C domains and this region forms two more β strands, C′ and C″. The function of V domains is to bind an antigen and the antigen-binding site is formed by three loops located between the B and C strands, the C′ and C″ strands, and the F and G strands, which are referred to as complementarity-determining regions (CDRs) I, II and III, respectively (Kabat et al. 1991). Therefore, the addition or deletion of C′ and C″ strands should be an important factor in determining the functions of V-related domains and of C-related domains. The V-related and C-related domains have been referred to as the V-set and the C1-set, respectively. Williams (1987) proposed one more category, the C2-set, since some proteins with characteristics of the Ig superfamily cannot be easily classified as members of the V-set or the C1-set. However, the classification with respect to the V-set and the C2-set, as well as the C1-set and C2-set, is arbitrary and, in some cases, it is difficult to classify proteins appropriately. Figure 2 shows typical examples of sequence alignments of proteins in the Ig superfamily. Since extensive alignments have been published (Williams and Barclay 1988, 1989; Hunkapiller and Hood 1989), this figure represents only part of the available data.

Classification of members of the Ig superfamily into the V-set, C1-set and C2-set has been based for the most part on sequence alignment (Williams 1987; Williams and Barclay 1988, 1989). In addition to those of antibodies, the three-dimensional structures of CD4 (Ryu et al. 1990; Wang et al. 1990), CD8 (Leahy et al. 1992), CD2 (Jones et al. 1992), β2-microglobulin (Becker and Reeke 1985), MHC class I and class II molecules (Bjorkman et al. 1987; Brown et al.

	B C		E F
Ig V(λ)	GETVTLTCRSSTGAVTTSNYANWVQEKPDH	——(23)——	LIGDKAALTITGAQT-EDEAIYFCALWYSNH
C(λ)	ENKATLVCTITDFYP-GVVTVDWKVDGTPV	——(15)——	KYMASSYLTLTARAW-ERHSSYSCQVTHEGH
TCR V(β)	GQEVTLRCKPISG----HNSLFWYRQTMMR	——(28)——	PNASFSTLKIQPSEP-RDSAVYFCASSFSTC
C(β)	KQKATLVCLARGFFP-DHVELSWWVNGKEV	——(16)——	SHCLSSRLRVSATFWHNPRNHFRCQVQFHGL
MHC H2K^b(α3)	EDKVTLRCWALGFYP-ADITLTWQLNGEEL	——(12)——	GDGTFQKWASVVVPL-GKEQYYTCHVYHQGL
IA^β(β2)	NHHNTLVCSVTDFYP-AKIKVRWFRNGQEE	——(12)——	GDWTFQVLVMLEMTP-HQGEVYTCHVEHPSL
NCAM(IV)	EDQITLTCEASGDP---IPSITWKTSTRNI	——(14)——	SHARVSSLTLKEIQY-TDAGEYVCTASNTIG

Fig. 2. Alignment of amino acid sequences of members of the Ig superfamily. *Bars* indicate missing amino acids. *Numbers in parentheses* indicate those of amino acid residues. The highly conserved amino acids are *boxed*. The data of amino acid sequences are taken from the following literature: *Ig V(λ)* and *C(λ)* of mouse, Bernard et al. (1978); *TCR V* of human, Yanagi et al. (1984); *TCR C* of mouse, Hedrick et al. (1984); *MHC H2K*^{b} (α3), Weiss et al. (1983); *MHC IA*^{β} (β2), Malissen et al. (1983)

1993), and MHC-related neonatal Fc receptor (Burmeister et al. 1994) are the only relevant structures reported to date. In the case of MHC molecules, however, the membrane-proximal domains form an Ig fold but the peptide-binding domains form a different structure (see below). The most typical characteristic of proteins that belong to the Ig superfamily seems to be two cysteine residues on the B and F strands, respectively, but they are not a prerequisite for formation of the Ig fold. Even in the case of the V domain of antibodies, the S–S bond is missing in one case but the antibody still retains antigen-binding capability (Rudikoff and Pumphrey 1986). The S–S bond should play a role in enhancing the stability of the Ig fold. A more important factor in formation of the Ig fold is a core structure formed by hydrophobic interactions among several hydrophobic amino acid residues that are located on the B, C, E and F strands (Lesk and Chothia 1982). In these β strands, hydrophobic and hydrophilic amino acids alternate, with the hydrophobic side chains pointing inwards, to form the interior of a sandwich between two β sheets, and with the hydrophilic side chains protruding outwards. Therefore, when amino acid sequences are aligned, the strong tendency toward conservation of these hydrophobic amino acids is apparent (Williams and Barclay 1988, 1989). In addition to two cysteine residues on the B and F strands, a tryptophan residue located on the C strand is conserved in many cases. This tryptophan residue is located at a key position with respect to formation of the Ig fold (Lesk and Chothia 1982; Amzel and Poljak 1979).

3 When Did Ig Appear in Evolution?

If we define an immune system as a self-defense system with the ability to recognize foreign substances specifically, only vertebrates can be said to possess a "true" immune system. It has been considered for many years that cyclostomes, such as the hagfish and lamprey, are the most primitive animals that produce antibodies. Since cyclostomes are the most primitive vertebrates, it would seem that all vertebrates produce antibodies. Antibodies can be defined as molecules with the following characteristics: (1) their production is induced by immunization with a particular antigen; (2) they bind the corresponding antigen specifically; and (3) a memory system exists for their production. Many classical studies using lamprey and hagfish seem to support the presence of antibodies in cyclostomes (for review, see Litman and Marchalonis 1982). For example, Marchalonis and Edelman (1968) reported that the sea lamprey *Petromyzon marinus*, one of the cyclostomes, produces a specific antibody in response to immunization with bacteriophage f2. Raison et al. (1978) showed that the hagfish, *Eptatretus stoutii*, also produces an antibody against group A streptococcal carbohydrate after prolonged immunization with a whole-cell preparation. Subsequently, hagfish antibodies were purified by three groups of investigators (Kobayashi et al. 1985; Hanley et al. 1990; Varner et al. 1991). Hanley et al. (1990) reported that the isolated antibody had high binding affinity for the streptococcal carbohydrate used for immunization. Litman and coworkers (Varner et al. 1991) reported partial amino acid sequences of a putative Ig isolated as a reduction-sensitive heterodimer from hagfish serum. From the physical properties, serological cross-reactivity with antisera against Ig of horned shark (*Heterodontus francisci*), and amino acid sequences of the isolated molecules, they suggested that the isolated protein was an Ig. However, Ishiguro et al. (1992) reported that the proteins that had been isolated as antibodies from hagfish do not resemble mammalian Ig but actually resemble complement C3. First, they determined the N-terminal amino acid sequence of the heavy (H) chains of a putative Ig that had been isolated from another species of hagfish, *Eptatretus burgeri*, by Kobayashi et al. (1985), and then they isolated cDNA clones that encoded this molecule. The amino acid sequence predicted from the nucleotide sequence of the cDNA indicated that this gene actually encodes proteins isolated as hagfish "antibodies" by various investigators. However, such proteins do not resemble mammalian Ig but possess certain characteristics common to the complement components C3, C4 and C5 of higher vertebrates (Belt et al. 1984; De Bruijn and Fey 1985; Wetsel et al. 1987). There is sufficient evidence to allow us to conclude that this protein corresponds to C3 since the gene isolated by Ishiguro et al. (1992) encoded the protein that had been functionally identified and isolated as hagfish C3 by Fujii et al. (1992). In cyclostomes, both in the lamprey and in the hagfish, the presence of complement C3 has been well documented (Nonaka et al. 1984; Fujii et al. 1992). Raison's group (Hanley et al. 1992) reached the same conclusion as that of Ishiguro et al. (1992). The conclusion that the proteins isolated as "antibodies" from hagfish correspond to complement C3, is true not only for hagfish proteins but also for lamprey proteins, as indicated in Fig. 3 (Marchalonis and Schluter 1989). The lamprey's "antibody" seems to be the same molecule as C3. It has been functionally identified and its amino acid sequence was determined by Nonaka and Takahashi (1992). Therefore, there is no evidence at present for true Ig molecules in cyclostomes.

In elasmobranchs, in contrast to the results for cyclostomes, Litman and coworkers succeeded in isolating genes for Ig from sharks (Hinds and Litman 1986). Initial success was achieved by cross-hybridization with a V_H probe specific for phosphoryl choline (PC) of mammals (Litman et al. 1985). The

Hagfish (Ishiguro et al. 1992)	
Hanley et al. (1992)	SKVLVIAPAAT
	SKVLVIAPAAT
	SQGEDFMIQE**T**
	SQGEDFMIQES
Varner et al. (1991)	STAYG**N**LAAIQ**P**EE
	STAYGLLAAIQHEE
	LNPI**SSP**LVV
	LNPILSSLVV
	LES**IQN**LPSPPASCQVS
	LESVCRLPSPPASCQVS
Lamprey (Nonaka and Takahashi 1992)	
Marchalonis (1991)	TFG**G**GV**EXV**L
	TFGDGVQKIL
	SYFP**E**SXGXN**T**YX**IP**
	AYFPQSWGWNKYKNS

Fig. 3. Comparison of amino acid sequences of hagfish and lamprey C3 with those of the putative hagfish "antibody" and lamprey "antibody", respectively. Sequences on the *upper lines* correspond to those of the putative "antibody" and those on the *lower lines* correspond to those of complement C3. Differences in amino acid residues are indicated in *bold letters*. The amino acid sequences of hagfish C3 and lamprey C3 were reported by Ishiguro et al. (1992) and Nonaka and Takahashi (1992), respectively. Partial amino acid sequences of the putative hagfish "antibody" have been reported by Raison and coworkers (Hanley et al. 1990) and Litman and coworkers (Varner et al. 1991). Marchalonis reported partial amino acid sequences of the putative lamprey "antibody" at the 5th ISDCI congress (1991)

nucleotide sequences of the shark and mammalian genes encoding these particular V_H genes have been strikingly well conserved. It has been clearly demonstrated that V regions of Ig are encoded by two or three split genes in germline genomes and that DNA rearrangements occur that result in combination of the split genes, such as the V, (D) and J genes, during B-cell ontogeny (Tonegawa 1983). The analysis of Ig genes in sharks indicates that the elasmobranchs have evolved a system for DNA rearrangements and that this system in elasmobranchs is essentially the same as that in mammals, including the 12- and 23-spacer signals which have been characterized as signals for the putative recombinase in mammals (Sakano et al. 1981). It has been suggested that the DNA recombination system that involves such spacer signals was introduced into immune systems at some point early in animal evolution, and this event made it possible for significant diversification of Ig and TCR gene systems to occur (Sakano et al. 1979).

In conclusion, elasmoblanchs are the most primitive animals that have so far been proven to possess a system for V(D)J rearrangements and to produce antibodies. The presence of antibodies in cyclostomes has not been proven.

4 When Did the TCR and MHC Appear in Evolution?

It has been argued that the occurrence of acute allograft rejection is characteristic of the presence of an MHC-TCR recognition system (Cohen and Borysenko 1970; Cohen 1979). Since neither cyclostomes nor elasmobranchs show acute allograft rejection, teleost fish have been considered to be the most primitive animals that possess this recognition system (Du Pasquier 1993). The presence of MHC molecules in teleost fish was suggested initially only by indirect results, such as acute allograft rejection (Hildemann 1970; Botham et al. 1980; Nakanishi 1987), a mixed lymphocyte reaction (Caspi and Avtalion 1984; Miller et al. 1986; Kaastrup et al. 1988) and antibody responses in vitro (Miller et al. 1985). More recently, Hashimoto et al. (1990) devised a method for isolating MHC genes directly from animal genomic DNA using the polymerase chain reaction (PCR). They synthesized two sets of degenerate oligonucleotides that corresponded to well-conserved sequences around two cysteine residues in membrane-proximal domains of MHC molecules as shown in Fig. 4 and then carried out PCR. The products amplified from carp genomic DNA appeared to correspond to parts of MHC class I and class IIβ genes. Using the amplified DNA segments as probes, they succeeded in isolating MHC genes from carp (Hashimoto et al. 1990). Their strategy is powerful for the direct isolation of Ig-related genes from mRNA and/or genomic DNA of various animals without a requirement for antiserum against a target protein (Hashimoto and Kurosawa 1991). Indeed, MHC genes have been successfully isolated from reptiles using this method (Grossberger and Parham 1992). When the sequences of primers are properly devised, not only genes for MHC, but also those for Ig and TCR can be isolated from various vertebrates. This strategy has recently been applied to isolation of MHC genes from sharks, which do not exhibit acute allograft rejection. Genes for MHC class I, class II α and β chains were successfully isolated by Hashimoto et al. (1992), Kasahara et al. (1992) and Bartl and Weissman (1994), respectively. Although, with respect to class I, only a portion of the gene that corresponds to the $\alpha3$ domain has been reported, several amino acids considered to be important for the interaction of the $\alpha3$ domain with the T-cell coreceptor CD8, as well as that with $\beta2$-microglobulin, appear to have been well conserved in the shark sequence (Hashimoto et al. 1992). Very recently, we isolated MHC class I genes from shark which encode three extracellular domains and whose sequences are highly homologous to those of mammalian classical MHC class I genes (Okamura, Kurosawa, Hashimoto, in prep). Therefore, it has been established conclusively that MHC molecules exist in elasmobranchs and that evolution of MHC molecules into classes I and II had already occurred at the elasmobranch level during evolution. This observation also implies the presence of TCR molecules in elasmobranchs.

The results described above indicate that the occurrence of acute allograft rejection is not a prerequisite for concluding that a TCR-MHC recognition system is present. Elasmobranchs have a thymus and a spleen and have the capacity to reject allografts, as observed in experiments with stingrays (Perey

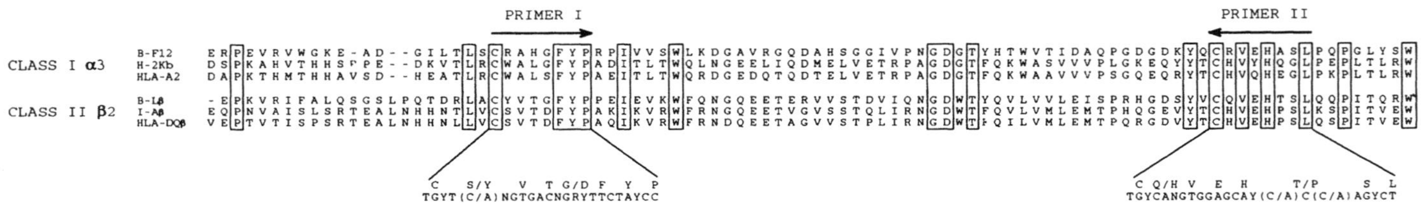

Fig. 4. Primers for PCR. The amino acid sequences of *class I α3* and *class II β2* domains of chicken, mouse, and human MHC molecules are shown. *B-F12* (Guillemot et al. 1988) and *B-Lβ* (Bourlet et al. 1988) of chicken, *H-2K*[b] (Weiss et al. 1983) and *I-Aβ* (Malissen et al. 1983) of mouse, and *HLA-A2* (Koller and Orr 1985) and *HLA-DQβ* (Larhammar et al. 1982) of man were chosen as representative examples. The amino acid residues that are identical in all these molecules are *boxed*. The locations of the two primers used for PCR are indicated by *arrows* and the DNA sequences of the primers and the corresponding amino acids are shown (in the case of primer II, the complementary sequence is shown). The sequences of the two primers were chosen mainly on the basis of class II *β* sequences. (Hashimoto et al. 1990)

et al. 1968) and horned sharks (Borysenko and Hildemann 1970). This reaction is not an example of acute allograft rejection since the initial rejection of skin allografts requires 4–8 weeks. However, second grafts are rejected in only 2–3 weeks, and third and fourth grafts are rejected still more rapidly in a process that resembles an acute type rejection. Apparently, cyclostomes possess neither a thymus nor a spleen (Du Pasquier 1993). However, they do have the capacity to reject allografts even though acute-type rejection is not observed (Hildemann and Thoenes 1969). Therefore, it remains likely that cyclostomes do have a primitive type of TCR-MHC recognition system. It has been argued that MHC genes for classes I and II could be derived from the same primordial gene (Kaufman et al. 1990), and it seems reasonable to postulate that two-domain molecules, such as class II, appeared earlier than three-domain molecules, such as class I (Kaufman et al. 1990). Therefore, the nature of MHC-like molecules in cyclostomes is an issue of considerable interest.

We can conclude that MHC class I and class II molecules exist in elasmobranchs and that TCR molecules should also exist in these animals, although the genes for TCR have been isolated in non-mammals only from amphibian and chickens to date (Tjoelker et al. 1990; Fellah et al. 1993; Göbel et al. 1994). After we submitted this review, Rast and Litman (1994) reported isolation of TCR genes from horned shark. In cyclostomes, although the presence of MHC and TCR molecules has not been demonstrated, it remains possible that they exist. If evidence is obtained from future experiments to show that cyclostomes have an MHC-TCR recognition system but do not produce antibodies, we will be able to deduce that TCR preceded Ig in evolution.

5 The Origin of Peptide-Binding Domains of MHC Molecules

The success of X-ray analysis of MHC molecules has revealed a striking feature of the peptide-binding domains (Bjorkman et al. 1987; Brown et al. 1993). The first and second domains of the MHC class I molecule and the first domains of MHC class II α and β chains form a flat platform that is composed of β sheets with two α helices on the platform. As shown schematically in Fig. 5, this structure is quite different from that of the Ig fold. Given not only the sequence similarity but also the structural similarity between molecules of class I and class II, the first domain of class I should correspond to the first domain of the class II α chain and the second domain of class I should correspond to the first domain of the class II β chain (Brown et al. 1988).

Two different hypotheses have been proposed to explain the possible origin of the peptide-binding domains of MHC molecules, as shown schematically in Fig. 6. According to one hypothesis, the peptide-binding domain may have originated from a protein that was totally different from members of the Ig superfamily and, as a candidate, heat-shock protein 70 (hsp70) has been proposed by two different groups (Flajnik et al. 1991; Rippmann et al. 1991). This hypothesis is based on the peptide-binding properties of hsp70, a comparison of its

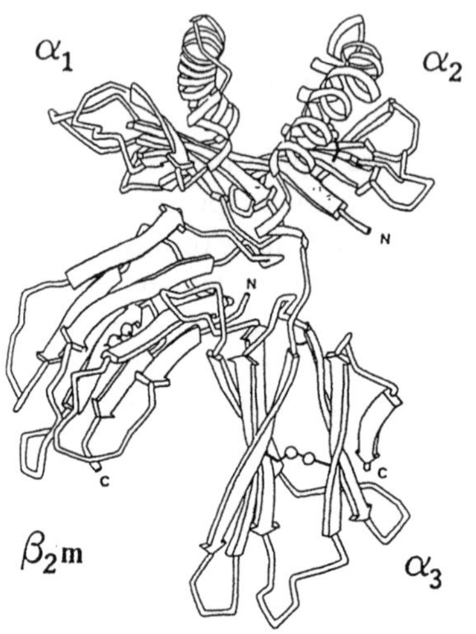

Fig. 5. Three dimensional structure of an MHC class I molecule (Bjorkman et al. 1987)

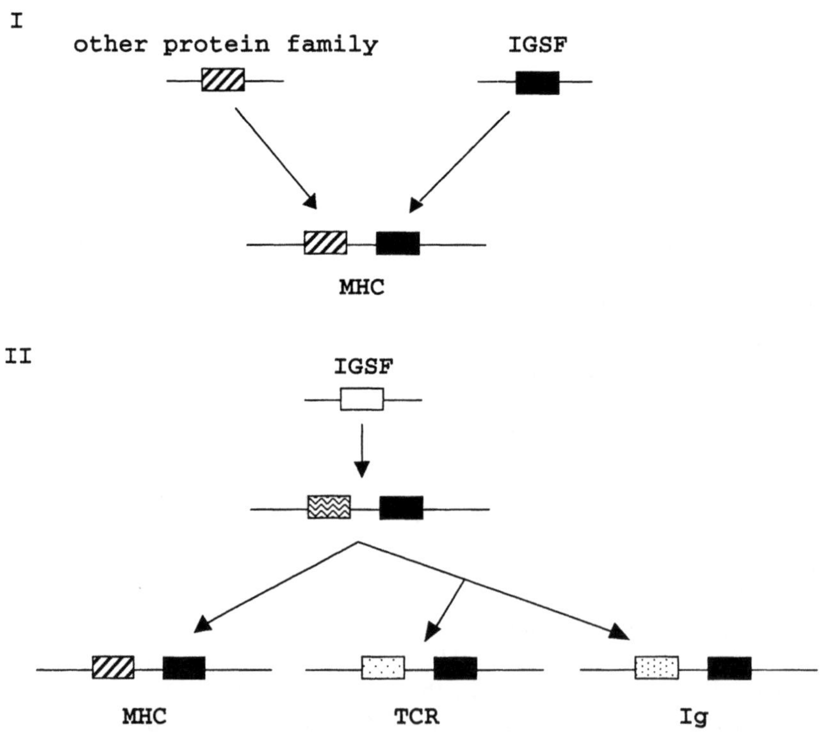

Fig. 6. Two models of the origin of peptide-binding domains of MHC. *IGSF* is an abbreviation for the Ig superfamily. One *box* corresponds to one domain. The details of the models are given in the text

amino acid sequence with those of MHC molecules, an estimation of the secondary structure of the putative peptide-binding region of hsp70, and the superimposition of this structure onto the three-dimensional structure of HLA. Although an essential role of hsp70-like molecules is to assist in folding and assembly of protein, these molecules also interact with $\gamma\delta$ TCR. Furthermore, genes for human hsp70 are located within the MHC gene loci. These observations seem to support the hypothesis that hsp70 served as a donor of the peptide-binding domains of MHC molecules. However, there are several proteins that adopt a structure composed of a platform of β-strands topped by α-helices, similar to the structure of the HLA molecules. For example, bovine platelet factor 4 (Charles et al. 1989) and interleukin 8 (Baldwin et al. 1991), both of which belong to the same family of protein, and bacteriophage coat protein MS2 (Valegård et al. 1990) include structures similar to that of peptide-binding domains of MHC. Bovine platelet factor 4 and interleukin 8 form dimers as do MHC class II molecules. There are differences, however, between these molecules and the HLA molecule in the angle between the β-strands and the α-helical axis (Charles et al. 1989). It seems probable that many molecules will be found, in the future, to have a structure composed of a platform of β-strands topped by α-helices. This structure could represent a stable three-dimensional configuration and many molecules, encoded by independent genes, could coincidentally adopt this structure. Determination of the three-dimensional structures of many further molecules, including the hsp70 and its peptide-binding region in particular, will be required if we are to clarify the relationship between these proteins and the peptide-binding domains of MHC molecules.

If the peptide-binding domains of MHC were derived from that of hsp70, how could primordial TCR molecules have been formed simultaneously? Without a large repertoire of TCR, peptide-binding MHC molecules would seem to be useless. A preferred model should include an explanation of the expansion of the TCR repertoire. Very recently, however, Srivastava et al. (1994) proposed the hypothesis that hsps, such as hsp70 and hsp90, constitute a relay line along which peptides, after generation in the cytosol by the action of proteases, are transferred from one hsp to another until they are finally accepted by MHC class I molecules in the endoplasmic reticulum, and that the binding of peptides by hsps constitutes a key step in the priming of cytotoxic T lymphocytes in vivo. According to this hypothesis, the peptide-binding regions of hsps and MHC are not only structurally similar to each other but are involved in related functions.

The second hypothesis relating to the peptide-binding domains of MHC molecules proposed that these domains were also derived from the Ig superfamily. Hashimoto et al. (1990) found sequence similarity between the amino acid sequence encoded by the first exon of the carp gene for the class II β chain (TLAIIβ-1) and the amino acid sequence of the extracellular domain of the CD8 β chain, in spite of the fact that this domain of the CD8 β chain is classified as a member of the V-set according to Williams' definition (Williams 1987). Figure 7 summarizes the comparison of amino acid sequences between the first domain of TLAIIβ-1 and the domains of members of the V-set. In the case of the rat CD8

β chain, approximately 50% of the amino acid residues are identical to those of TLAIIβ-1 in the middle region. It is of interest that this region corresponds to the S3′ and S4′ peptides in the proposed model for MHC molecules and the C′ and C″ peptides in the Ig fold. Furthermore, Fig. 7 shows that this region of TLAIIβ-1 also shares several identical residues with various members of the V-set. These residues include cysteine residues in the second domain of class I and the first domain of the class II β chain. On the basis of these observations, Hashimoto et al. (1990) proposed that peptide-binding domains with the characteristic α-helices of class I, as well as class II MHC antigens, also arose from members of the V-set.

The following observations appear to be consistent with the hypothesis that the peptide-binding domains of MHC molecules were derived from Ig-related domains. Most genes encoding molecules related to immune functions (which we call immune-related molecules, for simplicity), such as Ig, MHC, TCR, CD4 and CD8, are composed of exon and intron structures and one exon corresponds to one domain. Moreover, the boundary between each exon and intron is located between the first letter and the second letter in one codon. This regularity has been observed in the genes that encode immune-related molecules of the Ig superfamily in vertebrates, including the peptide-binding domains of the MHC. The hypothesis assumes the following two events. One required event is conversion from a β-sheet structure to an α-helical structure, and this conversion should have been achieved by changes in a limited number of amino acid residues. Such conversions have been shown experimentally to be possible in several proteins, such as scrapie prion proteins (Pan et al. 1993), the isolated photosystem II reaction center (He et al. 1991), and amyloid βA4 peptides (Kang et al. 1987; Hilbich et al. 1991). The second required event is that peptide-binding ability should have been gained by the amino acid sequence that originally formed β sheets. Analysis of the three-dimensional structure of PapD, a chaperone protein whose structure includes an Ig fold (see Sect. 7), that had been co-crystallized with an oligopeptide, indicated that the peptide bound in an extended conformation along one β strand in the cleft of the PapD protein (Kuehn et al. 1993). Thus, it is conceivable that Ig, TCR and MHC all originated from similar heterodimers composed of V and C domains. If this was the case, the major event in the creation of the peptide-binding domain from the V-set domain must have been conversion of a structure from β-sheets to α-helices. Deletion of the A peptide in the Ig fold could have been one possible driving factor for induction of such a conformational change.

In summary, two different molecules have been proposed as a possible origin of the peptide-binding domains of MHC molecules. One is hsp70, which has peptide-binding ability and whose structure is totally different from the Ig fold. The other is a V-set molecule. In either case, the hypothesis should be able to explain, in future experiments, the origin of the TCR, the origin of the polymorphism shown by MHC and the expansion of the TCR repertoire. Analysis of the MHC-like molecules that probably exist in cyclostomes and tunicates should help us to clarify the origins of the MHC and TCR.

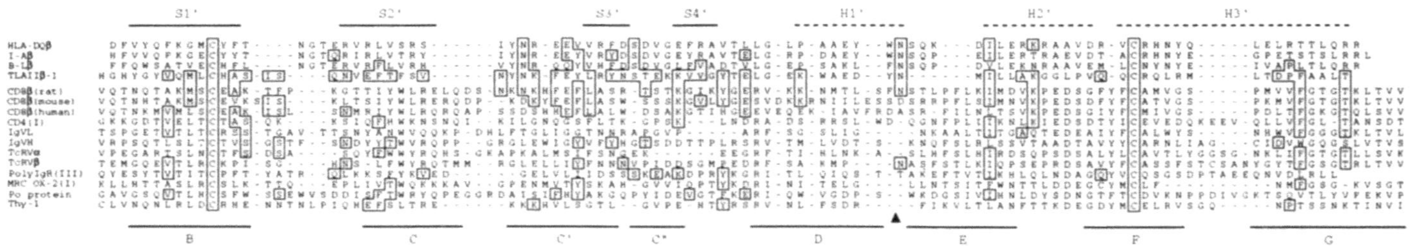

Fig. 7. Comparison of the amino acid sequence of exon 1 of *TLAIIβ-1* with V-set sequences of the Ig superfamily. V-set sequences are taken from Williams (1987), Williams and Barclay (1988) and Shiue et al. (1988) and the references cited therein and they are aligned as indicated by Williams and Barclay (1988). Alignment of the *last part* of the sequences (corresponding to the β-strand G in the immunoglobulin V domain) is in accordance with the alignments of Johnson and Williams (1986) and Williams and Gagnon (1982). No alignment was attempted in this region for the poly-immunoglobulin receptor and the Po protein. The amino acid sequences of MHC class II β1 domains of other species are shown *above* the sequence of *TLAIIβ-1*. Amino acid residues are *boxed* when they are shared by *TLAIIβ-1* and V-set member (s). *β-Strand* and *α-helices* observed in the α2 domain of MHC class I HLA-A2 are indicated by *solid lines* and *broken lines*, respectively, *above* the MHC class II sequences, as indicated by Bjorkman et al. (1987) and Brown et al. (1988), and the *β-strand* in the Ig V domain (Amzel and Poljak 1979) is indicated by *solid lines below* the C-set sequences. Comparison of the amino acid sequences between the V-set and the C-set suggested that several amino acids located *between D and E* are missing in the V-set (Williams and Barclay 1988). An *arrowhead* indicates this position. (Hashimoto et al. 1990)

6 The Ig Superfamily in Invertebrates

To date, more than ten kinds of protein in invertebrates have been proposed as members of the Ig superfamily. Figure 8 shows schematically those proteins that have been identified in insects and nematodes. The alignment of sequences clearly indicates that all of the proteins can definitively be classified as members of the Ig superfamily, as shown also by the alignment in Fig. 9.

Hemolin, isolated from the giant silk moth *Hyalophora cecropia*, has a unique characteristic (Sun et al. 1990). It has four Ig-fold domains. When silkworm larvae are injected with bacteria, larval hemocytes begin to synthesize hemolin in large amounts, and hemolin is one of the components that bind to the surface of the bacteria, taking part in formation of a protein complex that

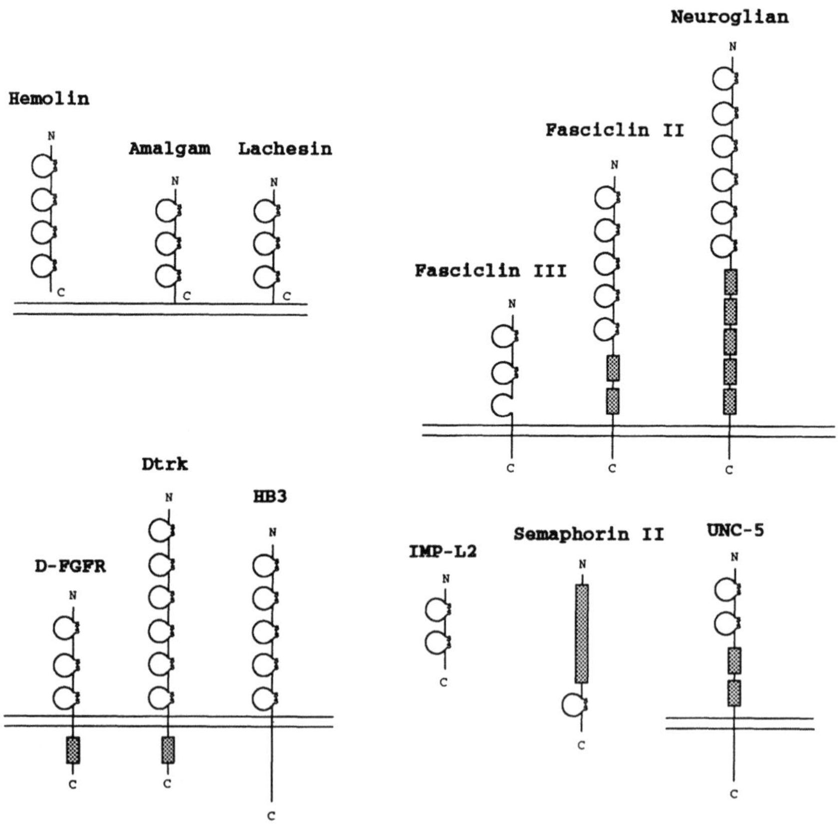

Fig. 8. Models for molecules in the Ig superfamily that have been identified in nonvertebrate phyla. Fasciclin III, fasciclin II, neuroglian, D-FGFR, Dtrk, HB3 and UNC-5 are membrane-bound proteins. IMP-L2 and semaphorin II are secreted proteins. Hemolin, amalgam and lachesin lack a transmembrane sequence, but they seem to be localized on the surface of the cells. Functional domains other than Ig fold domains are marked by *boxes*, and the details are given in the text. Twitchin of *C. elegans* and α-agglutinin of yeast that have also Ig domains are not included here

```
                           B                C                   E              F
Drosophila             ────────         ─────────           ─────────      ──────────
                    ▽    ▽ ▼ ▽   ▽ ▽ ▽    ▽ ▼ ▽    ▽            ▽ ▽       ▽ ▽ ▼ ▼ ▽ ▽  ▽
Hemolin           I NNPTVLECIIE--GNDQGV--KYSWKKDGKSY—(8)—LRKDEGSLVFLRP-QASDEGHYQCFAETPAGV
                 II GRPFQLDCVLP-NAYPKP---LITWKKRLSGA—(16)—TDGNLYFTIVTKE-DVSDIYKYVCTAKNAAVD
                III GDVTMIYCMYG--SNPMG---YPNYFKNGKDV—(11)—NRTSGKRLLFKTT-LPEDEGVYTCEVDNGVGK
                 IV GQDVTIPCKVT--GLPAP---NVVWSHNAKPL—(4)—ATVTDSGLVIKGV-KNGDKGYYGCRATNEHGD

Amalgam           I GDSVEFNCTVE--EVGQL---SVSWAKRPSES—(29)—TGSAIYTFRIQNI-EVSDMGPYECQVLVSATE
                 II GQNLELTCHAN--GFPKP---TISWAREHNAV—(5)—HLLAEPTLRIRSV-HRMDRGGYYCIAQNGEGQ
                III SHSAELECSVQ--GYPAP---TVVWHKNGVPL—(14)—SGTTTSVLRIDSV-GEEDFGDYYCNATNKLGH

Lachesin          I GGTVEFDCSVQ--YAKEY---NVLFLKTDSDP—(21)—PNSSTYKLQIKDI-QETDAGTYTCQVVISTVH
                 II GSEVQMECYAS--GYPTP---TITWRRENNAI—(5)—ATYVGNTLRIKSV-KKEDRGTYYCVADNGVSK
                III QYDMDLECHIE--AYPPP---AIVWTKDDIQL—(14)—DEYTDSTLRVITV-EKRQYGDYVCKATNRFGE

Fasciclin III     I YGRSINYCRIEIPGEQKVLNLSPEWSKTPGFT—(5)—LTAGQCGVSIERV-KASNNGQVKCSLGVEGEE
                 II GTEFRARCSVR-DGRPPA---NISWYIDNMPA—(18)—LSTSVQEIQWHLS-PEDSNRKLVCRSHHQTDR
                III EHTAIVNITIR--ASPQP---KIEWTIDGAIV—(18)—NDEYNVTLAIAGLTLEDTTKIYNLRASNELGL

Fasciclin II      I GKPLILTCRPT-VPEPSLVA-DLQWKDNRNNT—(17)—LPGESLALMITSL-SVEMGGKYYCTASYANTE
                 II GQDYVVMCEVK--ADPNP---TIDWLRNGDPI—(6)—YVVQTNGLLIRNV-QESDEGIYTCRAAVIETG
                III GKPFAANCTAR--GKPVP---EISWIRDATQL—(9)—VNPQTGLVTISSV-SQDDYGTYTCLAKNRAGV
                 IV TKEIAITCRAK--GRPAPAITFRRWGTQEEYT—(19)—RGESTGTLRISNA-ERSDDGLYQCIARNKGAD
                  V QRKANLSCLAM--GIPNA---TIEWHWNGRKI—(11)—GTGPRSDLIVHPV-TRQYYSGYKCIATNIHGT

Neuroglian        I DNPFIIECEAD--GQPEP---EYSWIKNGKKF—(10)—RQPGRGTLVITIP-KDEDRGHYQCFASNEFGT
                 II GEPFMLKCAAP-DGFPSP---TVNWMIQESID—(14)—PEGNLWFSNVTRE-DASSDFYYACSATSVFRS
                III GKRMELFCIYG--GTPLP---QTVWSKDGQRI—(7)—QGHYGKSLVIRQT-NFDDAGTYTCDVSNGVGN
                 IV DEEVVFECRAA--GVPEP---KISWIHNGKPI—(8)—RTVTDNTIRIINL-VKGDTGNYGCNATNSLGY
                  V GRNVTIKCRVN--GSPKP---LVKWLRASNWL—(5)—NVQANGDLEIQDV-TFSDAGKYTCYAQNKFGE
                 VI GQSATFRCNEA--HDDTLEI-EIDWWKDGQSI—(8)—VKTNDNSLTIAKT-MELDSGEYTCVARTRLDE

D-FGFR            I GDNVSIECLLE-DSALEP---KITWLHKGNAD—(24)—RMDEPQVLRLGNV-LMEDGGWYICIAENQVGR
                 II GNTVNLACPVY---GKA----NITWTKDKKPL—(6)—YVQKNWTLRFVEA-TSEDSGLYNCKVCNAWGC
                III NGSLVMKCTVY--SDLHP---TVSWKRVVLKN—(17)—VTNDSVVLTLRNV-TFDQEGWYTCLASSGLGR

Dtrk              I NESVKFECEST--DSYSEL--HYDWLHNGHRI—(5)—VHQIGSNLHIEAVRRTEDVGNYVCIATNLASG
                 II RNELLLKCHVE-GASGDLEPLEIEWYRNSEKL—(5)—VQLDQHRLIIRQP-GSEDDGLYRCTASNAAGR
                III VYQSSVKCLPR-LPRRKNQKMMESWDK-----—(0)—-------------------QTFLCRGKRGGAA
                 IV GEHTALTCLYELPDELKNQRIQLRWRKDGKLL—(31)—LHKQNGTLSFASI-IASDAGQYQCQLQLEAHA
                  V AVVAKVHCKAQ--GTPTP---QVQWVRDGENT—(6)—EVDANGTLIFRNV-NSEHRGNYTCLATNSQGQ
                 VI QGTVVMHCQAI--GDPKP---TIQWDKDLKYL—(10)—RFLENGTLEIRNV-QVEDEGSYGCTIGNSAGL

HB3               I GARVTLPCRVI---NKQG---TLQWTKDDFGL—(17)—DEEGDYSLDIYPV-MLDDDARYQCQVSPGPEG
                 II DRKVEIECVSV--GGKPAA--EITWIDGLGNV—(16)—RFTAKSVLRLTPK-KEHHNTNFSCQAQNTADR
                III HSQVRLECRAD--ANPSDV--RYRWFINDEPI—(1)—GGQKTEMVIRNVT-RKFHDAIVKCEVQNSVGK
                 IV GSVVSLTCEVD--SNPQP---EIVWIQHPSDR—(0)—VVGTSTNLTFSVS--NETAGRYYCKANVPGYA
                  V GDTARIECFAS--SVPRAR--HVSWTFNGQEI—(15)—PGGVKSTLIIRDS-QAYHYGKYNCTVVNDYGN

IMP-L2            I GATIEIVCEMM--GSQVP---SIQWVVGHLPR—(17)—AIVRVRSSHIIDH-VLSEARTYTCVGRTGSKT
                 II GSNIQLPCRVH--ARPRA---EITWLNNENKE—(7)—RVLANGDLLISEI-KWEDMGNYKCIARNVVGK


Semaphorin II       GQSVHLGCFVK-IPEVLKNE-QVTWYHHSKDK—(12)—IETTERGLVVVSV-NEADGGRYDCHLGGSLLC

C. elegans
UNC-5             I NKPLRLQCRAN-HATKIRYKCSSKWIDDSRIE—(13)—IDASVDISRIDVD-TSGHVDAFQCQCYASGDD
                 II GTTLQLPCQAP--ESDPKA--ELTWYKDGVVV—(6)—IRASDGSLIMSAA-RLSDSGNYTCEATNVANS

Twitchin (X26)      G****F*V*V*--G*P*P---*V*W*KNG**I-(5-10)-*******L*I**A-*R*DAG*YTI*A*N**G*
                         L I F            I    G   L                    V     S   KV V

S. cereivisiae
α-agglutinin        NNNVDLDCSSV-QVYSSNDF-NDWWFPQSYND—(28)—NALQSLPANVNTI-DHALEFQYTCLDTIANTT
```

Fig. 9. Alignments of amino acid sequences of members of the Ig superfamily in *Drosophila*, *C. elegans* and yeast. In the case of *twitchin*, only amino acid residues of the consensus sequence are indicated. Highly conserved and relatively well-conserved residues are marked by *filled triangles* and *open triangles*, respectively. Amino acid sequences: hemolin, Sun et al. (1990); amalgam, Seeger et al. (1988); lachesin, Karlstrom et al. (1993); fasciclin III, Snow et al. (1989); fasciclin II, Grenningloh et al. (1991); neuroglian, Biebar et al. (1989); D-FGFR, Glazer and Shilo (1991); Dtrk, Pulido et al. (1992); HB3, Ramos et al. (1993); IMP-L2, Garbe et al. (1993); semaphorin II, Kolodkin et al. (1993); UNC-5, Leung-Hagesteijn et al. (1992); twitchin, Benian et al. (1989); α-agglutinin, Lipke et al. (1989)

probably initiates a self-defense system. So far, hemolin is the only case in which a member of the Ig superfamily has been shown to be involved directly in the self-defense system of an invertebrate.

Amalgam and lachesin, isolated from *Drosophila*, appear to have a similar domain-structure (Seeger et al. 1988; Karlstrom et al. 1993). The amalgam protein of 333 amino acids consists of a signal sequence, three Ig-like domains and a short hydrophobic C-terminal region. The lachesin protein of 359 amino acids has essentially the same structure as amalgam. The N-terminal domains of both proteins are V-like and the other two domains are C-like. During embryogenesis, amalgam accumulates on the surface of various mesodermal and neural cells. The gene for amalgam is located between the *bcd* and *Dfd* genes in the *Antennapedia* complex in *Drosophila*. Lachesin is expressed on the surfaces of differentiating neural cells from the onset of neurogenesis in both the central and the peripheral nervous system (Karlstrom et al. 1993).

Cell adhesion molecules play a major role in regulating the outgrowth of neuronal growth cones and guiding their direction. In vertebrate nervous systems, many molecules have been reported to be involved in mediating outgrowth, guidance and fasciculation of axons. Many of these molecules are members of the Ig superfamily (Jessell 1988). In invertebrate nervous systems, several adhesion molecules have been identified and some of them have also been classified as members of the Ig superfamily. Fasciclin III of *Drosophila* has three Ig-fold domains, but its sequence has diverged considerably from those of amalgam and lachesin (Snow et al. 1989). Fasciclin III is capable of mediating adhesion in a homophilic manner. There are two kinds of combination of molecules that can mediate cell–cell interactions. One combination involves homophilic interaction, that is, one molecule recognizes another of the same type. Typical examples are the respective interactions of N-CAM (Edelman 1988) and several cadhelins (Takeichi 1990). The other combination involves heterophilic interaction, that is, different molecules recognize each other. Many of the surface antigens expressed on lymphocytes mediate this type of interaction (Springer 1994).

Fasciclin II and neuroglian seem to form a class, within the Ig superfamily, of proteins that function as neural recognition molecules (Harrelson and Goodman 1988; Biebar et al. 1989). Fasciclin II and neuroglian correspond to mammalian N-CAM and L1, respectively (Harrelson and Goodman 1988; Biebar et al. 1989). All of these proteins are expressed on a large subset of glial and neural cell bodies. Fasciclin II and N-CAM consist of five Ig domains and two fibronectin (Fn) type III domains, whereas neuroglian and L1 consist of six Ig domains and five Fn domains. The Ig domains in these molecules belong to the C2-set. These molecules appear to function in the selective fasciculation of specific neurons (Grenningloh et al. 1991) and the interactions between these molecules occur in a homophilic manner.

Proteins known as D-FGFR and Dtrk of *Drosophila* have unique characteristics since, in their cytoplasmic regions, they include tyrosine kinase domains (Klämbt et al. 1992; Pulido et al. 1992). D-FGFR has been isolated as a *Drosophila* homologue of the mammalian receptor for fibroblast growth factor (FGF; Glazer and Shilo 1991) and Dtrk has also been isolated as a *Drosophila* homologue of members of the trk family of mammalian receptors for neurotrophin that are tyrosine kinase-type receptors (Pulido et al. 1992). The D-FGFR protein consists of three Ig domains in its extracellular region and a kinase domain in its cytoplasmic region. The gene is expressed in a restricted set of tissues: the developmental tracheal system and the delaminating midline glial and neural cells. The presence of the D-FGFR protein is essential for the ability of migrating cells to recognize external guiding cues (Klämbt et al. 1992). The Dtrk protein consists of six Ig domains and a tyrosine kinase domain. Two of the Ig-fold domains, the third and the fourth, seem to belong to the V-set, and the other four domains belong to the C2-set. This protein is dynamically expressed during embryogenesis in *Drosophila* in several areas of the developing nervous system, including neurons and fasciculating axons (Pulido et al. 1992).

HB3 protein, encoded by the *irre-C* locus of *Drosophila*, is a transmembrane protein with five extracellular Ig domains and it is similar to the axonal surface glycoprotein DM-GRSP of the chicken (Ramos et al. 1993). A relatively long portion of this protein is apparently located in the cytoplasmic region. The gene for HB3 is expressed in the developing optic lobe and in the eye imaginal disc. HB3 protein has been shown to function in programed cell death (apoptosis) and in the process of axonal pathfinding (Ramos et al. 1993).

The gene for IMP-L2 of *Drosophila* has been identified as a gene that is induced by 20-hydroxyecdysone (Garbe et al. 1993). In embryos, IMP-L2 is first expressed at the cellular blastoderm stage and continues to be expressed throughout subsequent development. The characteristics of IMP-L2 are those of a secreted member of the Ig superfamily. One more secreted protein of *Drosophila*, semaphorin II, has been shown to contain an Ig-like domain (Kolodkin et al. 1993). Although semaphorins are encoded by a gene family, only semaphorin II has an Ig-like domain at the C-terminus. Semaphorin I (fasciclin IV) was originally identified in grasshopper as a molecule that functions in the guidance of two sensory growth cones (Kolodkin et al. 1992). Its homologues have been identified in *Drosophila*, human and chicken. Semaphorin I and II contain conserved semaphorin domains and semaphorin I has transmembrane domains; semaphorin II has no transmembrane domains.

In nematodes, two proteins have been classified as members of the Ig superfamily. The product of the *unc-5* gene is required for guiding pioneering axons and migrating cells along the body wall in *Caenorhabditis elegans*. The UNC-5 protein is a transmembrane protein of 919 amino acids. It has two Ig domains and two thrombospondin type 1 domains in its extracellular region and its large intracellular C-terminal region includes the SH3-like motif (Leung-Hagesteijn et al. 1992). It has been suggested that the SH3 domain mediates binding to microfilaments. The *unc-22* gene of *C. elegans* encodes a large muscle protein, called twitchin (Benian et al. 1989), which contains two repeating units, one of which is an Ig domain that is repeated 26 times, and the other of which is rich in proline residues and is repeated 31 times. Twitchin also contains a protein kinase domain and it is thought to be involved in the regulation of the activity of myosin.

As summarized here, eleven and two members of the Ig superfamily have been identified to date in insects and in nematodes, respectively. The numbers of such proteins should increase as a result of future experiments. After submitting this review, we noted that Shishido et al. (1993) had reported two FGF-receptor homologues of *Drosophila* belonging to the Ig superfamily. DFR1 contains two Ig domains and DFR2 contains five Ig domains. DFR2 turned out to be the same as D-FGFR that had been reported by Glazer and Shilo (1991). Although only three Ig-domain structures were indicated in Fig. 8, the whole molecule should contain five Ig domains. From the available data, we can conclude that the Ig superfamily existed before the divergence to arthropods (protostomes) and vertebrates (presumed derivatives of deuterostomes) during evolution.

7 The Ig Superfamily in Microorganisms

The presence of members of the Ig superfamily has been suggested even in microorganisms. The most plausible candidate to date is a protein in yeast, namely, α-agglutinin, a cell adhesion glycoprotein that is expressed on the walls of *Saccharomyces cerevisiae* α cells. As indicated in Fig. 9, it clearly shows characteristics of the Ig superfamily (Wojciechowicz et al. 1993). The Ig-like domain is located in the middle portion of the α-agglutinin protein and seems to play a role in binding to its ligand **a**-agglutinin, which is expressed by **a** cells. The fact that this member of the Ig superfamily found in a fungus is involved in the mating process is of considerable interest since, in the case of unicellular organisms, mating is the only phenomenon that requires interactions between cells.

Members of Ig superfamily may exist also in bacteria. Three-dimensional structural analysis of the chaperone protein PapD, which mediates the assembly of pili in *E. coli*, indicates that its polypeptide chain folds into two Ig-like domains (Holmgren and Bränden 1989; Holmgren et al. 1992). Although Holmgren and Bränden (1989) suggested initially that horizontal gene transfer may have occurred, they argued more recently that PapD could be the protein of an ancient gene (Holmgren et al. 1992). Juy et al. (1992) reported one further example of an Ig-fold domain in a protein from a microorganism. The N-terminal domain of endoglucanase CelD from *Clostridium thermocellum* has an Ig-like domain although the amino acid sequence of this domain is different from that of members of the Ig superfamily.

Ig-like domains have been identified in proteins encoded by animal viruses such as Epstein–Barr virus (EBV; Wang et al. 1989; Cashman and Pouliot 1990). However, since viral genomes often include genes that were originally encoded by the host genome, such observations do not necessarily mean that the viral genomes evolved to include an Ig-like gene. A similar situation has been observed in the case of cytomegalovirus. Beck and Barrell (1988) reported that human cytomegalovirus encodes a glycoprotein homologous to MHC class I antigens.

If α-agglutinin is a true member of the Ig superfamily, the members of the Ig superfamily must have existed prior to the divergence of fungi and animals and played a role in cell–cell recognition. Even so, it is very difficult to make the argument that all the genes for members of the Ig superfamily found in animals and fungi are derived from the same primordial gene. Rather, it seems more plausible that the structure of the Ig fold represents one of the most stable forms of proteins that are to be involved in protein–protein interactions. Very recently, in fact, Overduin et al. (1995) reported that the E-cadealin domain responsible for selective cell adhesion shows structural similarities with the immunoglobulin fold. Therefore, during evolution, in the case of proteins involved in such interactions, some proteins with this type of structure may have been selected independently.

8 The Immune System vs the Nervous System

As described above, many proteins belonging to the Ig superfamily are involved in the immune systems and in the nervous systems of vertebrates and invertebrates. A similarity between the immune system and the nervous system was first pointed out by Jerne in 1974, as follows (Jerne 1974):

"These two systems stand out among all the organs of our body by their ability to respond adequately to an enormous variety of signals. The cells of both systems can receive as well as transmit signals and the signals can be either excitatory or inhibitory. Both systems form a network either via neurons or idiotype-anti-idiotype interactions of V regions of Ig. The network resides in the ability of these elements to recognize as well as to be recognized. The modulation of the network by foreign signals represents its adaptation to the outside world. Both systems learn from experience and build up a memory."

Finally, Jerne predicted that these striking analogies between the two systems might have resulted from the similarities in the sets of genes that govern their expression and regulation.

Data in support of this statement have been provided by recent collections and classifications of genes involved in both immune systems and nervous systems. Molecular components of the immune system include Ig, MHC, TCR, complement, lymphokines and their receptors, as well as many surface antigens that are expressed on lymphocytes and macrophages. Many of the membrane-bound molecules, including Ig, belong to the Ig superfamily (Williams and Barclay 1988, 1989). Several molecules that can be classified as members of the Ig superfamily have been detected in the nervous system. The first examples were N-CAM (Cunningham et al. 1987), myelin-associated glycoprotein (MAG; Lai et al. 1987) and the major glycoprotein of peripheral myelin (Po; Lemke and Axel 1985). Many other proteins from vertebrates, with homologues in *Drosophila* as described above, have been shown to belong to the Ig superfamily. Among them, myelin/oligodendrocyte glycoprotein (MOG) is particularly noteworthy (Gardinier et al. 1992). MOG is thought to mediate the axon–glial adhesion that precedes myelination, and the gene for MOG is located close to the MHC gene cluster in mice and humans (Pham-Dinh et al. 1993). The amino acid sequence of

MOG is quite similar (41%) to that of the B–G antigen of chicken, which is presumably involved in the immune system (Kaufman and Salomonsen 1992). Observations of this type suggest a correlation between the nervous system and the immune system.

Several cytokines, such as interleukin IL1, IL2, IL3, IL6, IL8, IL11, interferon γ, tumor necrosis factor (TNF) and leukemia inhibitory factor (LIF), which were originally isolated as lymphokines in the immune system, have been shown also to play roles in the nervous system (Bazan 1991; Hall and Rao 1992; Rot 1992; Patterson and Nawa 1993; Steinman 1993). Conversely, neurotrophins, such as nervous growth factor (NGF), which were originally isolated as neurotrophic factors, have been shown to function in the immune system (Morgan et al. 1989). The presence of the same or similar cytokines in the immune system and in the nervous system suggests that receptors for such cytokines exist in both systems. Since the two systems penetrate most tissues of the body, while being kept separate from each other by, in particular, the blood–brain barrier, secretions of cytokines in one system may not disturb the regulation of the other system, and each system may operate independently. It has been shown that HT7 protein, which seems to be a receptor involved in cell-surface recognition at the blood–brain barrier, can be classified as a member of the Ig superfamily (Seulberger et al. 1990).

When we consider the emergence and subsequent evolution of the nervous system and the immune system, we must bear in mind that the nervous system and the circulatory system, which includes the immune system, are the only systems that penetrate all other tissues of the body. All organisms require a self-defense system and the immune system represents the most efficient self-defense system against pathogens. The nervous system appeared in animals in the phylum Cnidaria. Since the immune and nervous systems are separated by the blood–brain barrier, they could have evolved separately without influencing each other. At their initial emergence, however, they were presumably not separated from each other. Members of the Ig superfamily, other adhesion molecules, and cytokines and their receptors, which were already available at that time, became involved in the respective biological systems. After the blood–brain barrier had been established, the evolution of both systems may have been accelerated. The similarities observed between the immune system and the nervous system may reflect not only the versatility of the molecules involved but also similarities in the evolutional history of the two systems.

9 Strategy for the Development of Self-Defense Systems in the Animal Kingdom

All living organisms are constantly exposed to the selective pressures that are exerted by pathogens, such as bacteria, viruses and parasites. In primitive animals, two kinds of self-defense system seem to have developed. One system involves direct killing of microorganisms by secretion of substances that are toxic for pathogens, such as attacins, cecropins and lysozyme (Boman and Hultmark 1987). These proteins are produced by various animals and their synthesis and appropriate secretion have been shown to be induced upon infection by microorganisms. The other self-defense system involves phagocytosis and the digestion of foreign particles by macrophages. It has been shown that the Porifera (sponges), which are the most primitive metazoans, have macrophage-like cells although the functions of such cells have not been fully characterized (Simpso 1984; Cooper et al. 1992).

As described in Section 3, the proteins isolated as antibodies from cyclostomes correspond to complement component C3. A complement system consists of plasma and cell-surface proteins that play an important role in amplification of the recognition and the elimination of foreign materials (Liszewski and Atkinson 1993). This process occurs through an activation cascade that results in the formation of bioreactive fragments, by proteolytic cleavage, and the assembly of a protein complex that is capable of lysing cells. Two pathways have been shown to trigger the activation cascade. The classical pathway of complement is initiated by the formation of an antigen–antibody complex. Initiation of the alternative complement pathway does not involve an antibody. C3 plays an essential role in both the classical and the alternative pathway of complement by interacting with various other components of plasma and the cell surface. In cyclostomes, the presence of the alternative pathway has been definitively demonstrated (Nonaka et al. 1984; Fujii et al. 1992). α2-Macroglobulin has been considered to be a prototype molecule of complement C3, C4 and C5 and it functions to inhibit proteases (Sottrup-Jensen et al. 1985; Adinolfi 1993). When pathogens secrete proteases that may destroy components of the host, α2-macroglobulin envelops them and then macrophages that express receptors for α2-macroglobulin efficiently incorporate them. This phenomenon is similar to the opsonin function performed by either complement or antibody (Armstrong et al. 1993). The function of C3 in cyclostomes could also be that of an opsonin. Thus, the efficiency of phagocytic uptake of pathogens could have been increased upon the emergence of C3 which acts as an opsonin.

The appearance of C3 seems to have occurred earlier than that of antibodies during evolution. In higher vertebrates, the number of antibodies is huge, and each antibody specifically binds to a different antigen. Without a huge repertoire, how could antibodies have played a role as a defense system? Initially, the number of Ig molecules must have been very small.

A large repertoire of the Ig molecules is achieved by DNA rearrangements in Ig gene loci. A system for rearrangements of DNA has been shown to be involved only in the generation of genes for Ig and TCR during an animal's development. It has been suggested that an insertion segment containing 12- and 23-spacer signals was inserted into the primordial V gene and that DNA recombinase operated to combine the separated DNA segments during somatic processes (Sakano et al. 1979). Introduction of the DNA rearrangement system into the system of Ig and TCR genes may have allowed the sequences of V regions to

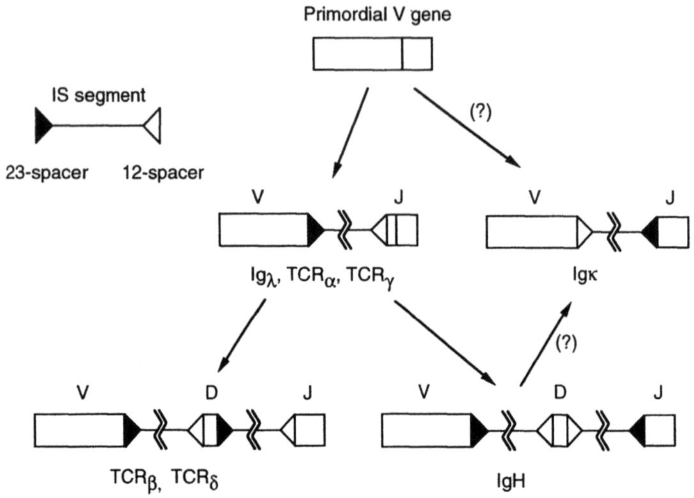

Fig. 10. Insertion model to explain the origin of the DNA rearrangement system at the *Ig* and *TCR* gene loci. A putative insertion (*IS*) segment with 12-spacer and 23-spacer signal sequences for DNA rearrangement at both ends was inserted into a gene that encoded a primordial *V* gene, with resultant formation of *V* and *J* genes. Insertion of the *IS* segment into the 5′ end region of *J* genes created one more gene, the *D* gene. Since only Igκ has an *IS* segment with opposite polarity, there are two possibilities: *Vκ* and *Jκ* genes could have been directly created; *Vκ* gene could have been created from V_H and *D* via V_HD joining

become highly diverged. Figure 10 summarizes the distribution of the signals in Ig and TCR genes that are found in contemporary animals. As judged from the consistency of the position of insertions in the V genes of Ig and TCR, DNA rearrangement mechanisms could have been introduced into these gene systems before Ig and TCR diverged during evolution. Location of 12- and 23-spacer signals at the Igκ gene loci is the opposite of those in the other gene systems. If the hypothesis that an insertion segment containing 12- and 23-spacer signals was inserted into the primordial V gene is correct, Igκ might be derived from IgH via a V_HD joining. If we assume that the MHC–TCR recognition system appeared first and that Ig genes emerged from parts of TCR genes, how could Ig have become involved in a self-defense system that already existed?

The number of genes for Ig in elasmobranchs has been shown to be large, but the degree of divergence of amino acid sequences of V genes of the Ig analyzed to date appears to be low (Kokubo et al. 1988). Although more extensive analyses are required to obtain a complete picture of the V gene repertoire in elasmobranchs, it seems probable that the antigens that can be recognized by antibodies in elasmobranchs are quite restricted (Ohno 1990). The fact that the sequence of V_H genes of the shark is similar to that of the gene for anti-PC antibody in mammals may reflect a specific position of the PC antigen, with respect to antigens recognized by antibodies in lower vertebrates, since an anti-PC antibody can recognize more than half of all bacteria (Ohno 1990). The antigen specificity of antibodies encoded by Ig genes in elasmobranchs should be examined in future experiments.

DNA rearrangements, such as the V-(D)-J joinings observed in mammals, play two roles in the production of Ig molecules. One is clearly to create the diversity of V regions of antibodies, whereas the other role relates to clonality of antibody-producing cells. Each respective antibody-producing cell should produce only one kind of antibody that responds to stimulation by one specific antigen. A situation wherein one cell produces multiple kinds of antibody would be dangerous to the host in some cases. Production of only one kind of antibody by each antibody-producing cell can be achieved in higher vertebrates since the number of genes for the C region is essentially only one. With regard to the clonality of antibody-producing cells, Ig genes in elasmobranchs do not seem to be organized in an appropriate way. For example, one unit, composed of V, (D), J and C genes, is repeated many times. Moreover, in some cases, DNA rearrangements seem already to have occurred in the genome of germ cells (Kokubo et al. 1988). Therefore, it seems to be impossible for elasmobranchs to retain the clonality of antibody-producing cells.

When an ancient antibody first appeared in the animal kingdom, the number of antigens that could be recognized by this antibody must have been restricted. If we assume that such antigens were restricted exclusively to molecules expressed on the surfaces of pathogens and that the main function of the antibody was that of an opsonin, the nature of phagocytic uptake of pathogens would have been changed to include antigen-specific recognition. Diversification of V regions of antibodies might have accelerated an increase in the efficiency of phagocytic uptake. Simultaneously, however, such diversification might have increased a potential risk, namely, production of anti-self antibodies. At the emergence of antibodies, however, it was not necessary for the production of antibodies to be regulated in terms of a system for discrimination between self and nonself.

In addition to Ig and TCR gene loci, Greenberg et al (1995) reported a new antigen receptor family (NAR) that undergoes rearrangement and extensive somatic diversification in sharks. Although ligands for NAR have not been revealed, the presence of NAR homologues in other vertebrates is an issue of great interest.

10 The Origin of Polymorphic Molecules: a Hypothesis

In contemporary higher vertebrates, self–nonself discrimination in the immune systems is mainly achieved by MHC and TCR molecules, as well as by selection of T lymphocytes in the thymus. MHC molecules are highly polymorphic proteins that bind various oligopeptides while TCR molecules form an enormously large repertoire. In ordinary ligand–receptor recognition systems, after the duplication of genes that encode both ligand and receptor molecules had occurred during evolution, changes in amino acid residues could have been introduced and fixed only when an interaction between the two molecules was able to occur. This mode of evolution of genes is thought to be diversification of the recognition systems. By contrast, in the case of the MHC–TCR recognition system, changes in amino acids introduced into the sequence of MHC molecules may have been recognized by some TCR molecules that had previously multiplied and diverged. Thus, the presence of a large repertoire of TCR and the presence of a selection system seem to have been required for the occurrence of polymorphism in MHC molecules. At the time when MHC molecules started to become polymorphic molecules, was the size of the TCR repertoire sufficiently large to allow recognition of the polymorphism of MHC molecules? Clonality of cells that express TCR also appears to be indispensable for selection of cells. In the case of T cells, introduction of a DNA rearrangement system at TCR gene loci must have been important if the respective T cells were to exhibit clonality. If we assume that the insertion hypothesis described in Sect. 9 is correct with respect to the establishment of DNA rearrangement systems, a primordial gene, which was a prototype of a gene for TCR before the immune system gained its DNA rearrangement system, must have existed. Therefore, the question to be answered is whether or not MHC molecules were polymorphic before TCR gained a DNA rearrangement system.

The presence of polymorphic molecules and the presence of a recognition system have been observed not only in the MHC–TCR systems of higher vertebrates but also in other biological systems. Although allorecognition observed in vertebrates is an artificial example that reveals the capacity for self–nonself recognition, the phenomenon of allorecognition has been observed in almost all metazoans including sponges (Hildemann 1977; Hildemann et al. 1979). To achieve allorecognition, an animal should express polymorphic determinants on the surface of its relevant cells and, simultaneously, it should have systems for recognition of the polymorphic molecules. Oka and Watanabe (1957) reported that, in tunicates, colony specificity is under the control of a single locus with multiple alleles (*Fu/HC* locus). Polymorphic proteins should be encoded at this *Fu/HC* locus. Individual tunicates showed heterozygosity at the *Fu/HC* locus and fertilization of an egg by a sperm occurred only when the respective genotype of each at this locus was different. Colonies differing from each other by one haplotype fused with each other, whereas if they differed at the level of the two alleles, they did not fuse. Two important conclusions were deduced from these results. Polymorphic molecules encoded at the *Fu/HC* locus are involved in two different biological phenomena, fertilization and fusion of colonies; they seem to function in opposite ways in the two phenomena. During fertilization, "differences" (nonself) seem to be essential while for fusion, "sameness" (self) is more important. Since, in this allorecognition system, one-haplotype-difference combinations were compatible for fusion, this system seems to be different from the MHC–TCR system in vertebrates where both haplotypes must be the same if they are to be compatible (Scofield et al. 1982). Weissman et al. (1990) reported subsequently a phenomenon: after fusion of two individuals that had a one-haplotype difference, one became dominant as a result of resorption of the other.

Therefore, even tunicates seem to have a system that functions for recognition not only of sameness but also of difference.

What could be the characteristics of the molecules involved in self-incompatibility and, in particular, in fertilization in animals? If the step at which self-fertilization is prevented is the adsorption of a sperm to the egg, the relevant molecules should be expressed on the surface of the egg and sperm. In such a case, it is unlikely that the molecules form a large repertoire and that a selection system exists during the process of maturation of the egg and sperm. According to the observations of Oka and Watanabe (1957), the same recognition system is involved in both fertilization and fusion, although in different ways. The molecules involved do not have to be different from each other as long as they can recognize each other either homophilically or heterophilically. Thus, we suggest the following scenario as an explanation of the emergence of polymorphism in the *Fu/HC* gene product, as shown schematically in Fig. 11. When multicellular organisms, such as sponges, first appeared, adhesion molecules played a role in cell–cell interactions. One of the genes that encoded such molecules was duplicated and mutations were introduced into the duplicated genes. Mutations were accepted when both molecules were able to interact with each other. The nature of the recognition was changed from homophilic to heterophilic. Such changes are in no way different from the diversification of a non-polymorphic recognition system. This early recognition system functioned originally in cell adhesion but then became involved in the process of fertilization. The most important aspect of our scenario for explanation of the emergence of a polymorphic molecule is that the genes should have been closely linked on a chromosome and were not separated during meiosis. We also have to assume that interactions between these molecules gave positive signals with respect to cell adhesion in an individual organism and with respect to fusion between two individuals, while the same interaction gave negative signals for fertilization. When the molecules expressed on an egg and a sperm cannot interact with each other, mating becomes possible. There is no direct evidence that polymorphic molecules emerged to prevent self-fertilization. However, a mechanism for prevention of self-fertilization has been widely observed not only in the animal kingdom but also in the

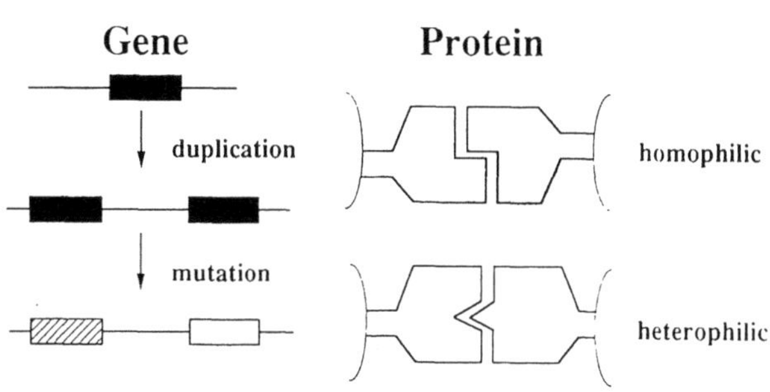

Fig. 11. Model to explain the origin of polymorphic molecules by concerted evolution. The details of the model are given in the text

plant kingdom and even in fungi (Herskowitz 1989). It is likely that organisms that gain a system for self-incompatibility gain an advantage in terms of maintenance of their respective species. Referring the observations of Oka and Watanabe, Burnet (1971) pointed out the similarities among three biological phenomena, namely, the immune system, compatibility in colonial tunicates and self-incompatibility in flowering plants. In these three systems, self–nonself recognition occurs. To effect this recognition, the presence of polymorphic molecules and their recognition systems is required.

Recent progress in research into the *S* (self-incompatible) loci in plants has suggested that the organization of genes, as predicted by our scenario, might be possible. Although the proteins encoded at the *S* loci in plants are totally different from those of the Ig superfamily, Stein et al. (1991) identified two genes at the *S* loci in *Brassica oleracea*, the gene for S receptor kinase (SRK) and the gene for *S*-locus glycoprotein (SLG); both genes are highly polymorphic. For example, the amino acid sequences of SLG and SRK proteins are found to be only 68% and 67% identical, respectively, when two individuals with different genotypes are compared. By contrast, the amino acid sequences of SLG and SRK from one individual can be as much as 90% identical. These results indicate that the amino acid sequences of SLG and SRK diverged rapidly but both sequences retained significant similarity in each respective plant. Stein et al. (1991) referred to this mechanism as concerted evolution. Although there is no direct evidence to date to show that SLG and SRK are really involved in self-incompatibility, the organization of these genes and the nature of the changes in amino acid sequences do not rule out this possibility.

We do not argue at present that the allorecognition system observed in lower animals should be thought of as a prototype of the MHC–TCR recognition system. It is quite possible that only two molecules, without a large repertoire, are involved in the former type of system. By contrast, TCR forms a large repertoire. In mammals, responders or non-responders to certain antigens, as well as high responders or low responders are correlated with various MHC genotypes. From such observations it has been argued that, since MHC molecules are polymorphic, animals belonging to one species can survive and procreate as a species. The presence of parasites may have been one of the driving forces that increased the polymorphism of MHC molecules (Klein et al. 1993). This possibility could hold true for contemporary animals. The polymorphism observed in MHC molecules seems to have become possible after the size of the repertoire of TCR became sufficiently large. The appearance of DNA rearrangement in immune systems might have occurred either at the cyclostome level or at the elasmobranch level. Polymorphic molecules encoded at *Fu/HC* loci in tunicates function primarily to prevent self-fertilization. Once males and females had become distinctly separate, such a system became non-essential. The following alternative possibilities must be examined. It is possible that the prototype molecules for MHC–TCR differed from molecules encoded at *Fu/HC* loci and were adhesion molecules that did not show polymorphism. After a DNA rearrangement system had been introduced into the primordial TCR gene, polymorphism started to appear in MHC molecules. As a second possibility, the genes at *Fu/HC* loci in tunicates were direct ancestors of MHC and TCR genes. If such was the case, MHC molecules started to be polymorphic from their emergence. Analysis of *Fu/HC* loci at the gene and the protein level should help us resolve these possibilities.

11 The Ig Superfamily in Perspective

In this review we described various characteristics of proteins that have been classified as members of the Ig superfamily. They are involved in surface-recognition events and in protein–protein interactions. The Ig fold probably has great advantages for a protein that is to play a role in protein–protein recognition. During evolution, once a new gene system had become established, multiplication and diversification of the system seem to have occurred frequently. As a result, multiplied and diverged systems started to be utilized in biological phenomena different from the original system in which the prototype system had been utilized (Yayon et al. 1992). Several examples have been reported of such situations, as follows.

Many *MHC* genes are encoded at the *MHC* loci. These genes include a few polymorphic, classical class I genes, as well as many nonclassical class I genes with low polymorphism. The *CD1* family is another group of class I-related genes that are not linked to the *MHC* gene loci (Calabi and Milstein 1986; Calabi et al. 1989). An Fc receptor for Ig in the neonatal rat was classified as a class I-related molecule (Simister and Mostov 1989). Recently, Okamura et al. (1993) analyzed extensively the genes that encode MHC class I molecules in cyprinid fishes and they demonstrated the active expansion of these genes in teleost fish. Very recently, we identified a human gene, MR1, outside the human MHC related to classical HLA class I genes (Hashimoto et al. 1995). Analysis of the biological roles of these molecules will allow us to identify functional varieties of MHC-like molecules.

Anderson et al. (1993) reported yet another relevant example. A fragment derived from complement component C3 and its receptor have been shown to be involved in the interaction between the oocyte and sperm. They suggested that regulated gamete-induced generation of C3 fragments and the binding of these fragments by selectively expressed receptors on the sperm and oocytes might be an initial step in the interaction of gametes, leading to membrane fusion and fertilization.

Proteins of the Ig superfamily have been identified mainly in the immune systems, but also in the nervous systems, of various invertebrates and vertebrates. The reason that they have been identified in these systems and not in other biological systems could be due to the history of the biological sciences. An Ig superfamily was originally proposed on the basis of the observations of proteins involved in the immune system. Moreover, the nerve cells and lymphocytes have been most extensively analyzed, among various biological systems, in terms of their surface proteins. As described in Sections 6 and 7, the Ig superfamily existed before the divergence to arthropods and vertebrates during evolution, and members of this superfamily probably exist even in yeasts and possibly in bacteria. If molecules of the Ig superfamily have such an ancient lineage, it is possible that they are involved in various, as yet unidentified, biological phenomena in contemporary animals. Future experiments with more varied kinds of tissue and analysis of additional biological phenomena should reveal new frontiers with respect to the significance of the Ig superfamily.

References

Adinolfi M (1993) Ontogeny and phylogeny of complement. In: Cooper EL, Nisbet-Brown E (eds) Developmental immunology. Oxford Univ Press, New York, pp 290–314

Alzari PM, Lascombe MB, Poljak RJ (1988) Three-dimensional structure of antibodies. Annu Rev Immunol 6: 555–580

Amzel LM, Poljak RJ (1979) Three dimensional structure of immunoglobulins. Annu Rev Biochem 48: 961–997

Anderson DJ, Abbott AF, Jack RM (1993) The role of complement component C3b and its receptor in sperm-oocyte interaction. Proc Natl Acad Sci USA 90: 10051–10055

Armstrong, PB, Armstrong MT, Quigley JP (1993) Involvement of α_2-macroglobulin and C-reactive protein in a complement-like hemolytic system in the arthropod, *Limulus polyphemus*. Mol Immunol 30: 929–934

Baldwin ET, Weber IT, Charles RS, Xuan JC, Appella E, Yamada M, Matsushima K, Edwards BFP, Clore GM, Gronenborn AM, Wlodawer A (1991) Crystal structure of interleukin 8: symbiosis of NMR and crystallography. Proc Natl Acad Sci USA 88: 502–506

Bartl S, Weissman IL (1994) Isolation and characterization of major histocompatibility complex class IIB genes from the nurse shark. Proc Natl Acad Sci USA 91: 262–266

Bazan JF (1991) Neuropoietic cytokines in the hematopoietic fold. Neuron 7: 197–208

Beck S, Barrell BG (1988) Human cytomegalovirus encodes a glycoprotein homologous to MHC class-I antigens. Nature 313: 269–272

Becker JW, Reeke NR, Jr (1985) Three dimensional structure of β2-microglobulin. Proc Natl Acad Sci USA 82: 4225–4229

Belt KT, Carroll MC, Porter RR (1984) The structural basis of the multiple forms of human complement component C4. Cell 36: 907–914

Benian GM, Kiff JE, Neckelmann N, Moerman DG, Waterston RH (1989) Sequence of an unusually large protein implicated in regulation of myosin activity in *C. elegans*. Nature 342: 45–50

Bernard O, Hozumi N, Tonegawa S (1978) Sequences of mouse immunoglobulin light chain genes before and after somatic changes. Cell 15: 1133–1144

Biebar AJ, Snow PM, Hortsch M, Patel NH, Jacobs JR, Traquina ZR, Schilling J, Goodman CS (1989) *Drosophila* neuroglian: a member of the immunoglobulin superfamily with extensive homology to the vertebrate neural adhesion molecule L1. Cell 59: 447–460

Bjorkman PJ, Saper MA, Samraoui B, Bennett WS, Strominger JL, Wiley DC (1987) Structure of the human class I histocompatibility antigen, HLA-A2. Nature 329: 506–512

Boman HG, Hultmark D (1987) Cell-free immunity in insects. Annu Rev Microbiol 41: 103–126

Borysenko M, Hildemann WH (1970) Reactions to skin allografts in the horn shark, *Heterodontis francisci*. Transplantation 10: 545–551

Botham JW, Grace MF, Manning MJH (1980) Ontogeny of first set and second set alloimmune reactivity in fishes. In: Manning MJ (ed) Phylogeny of immunological memory. Elsevier, Amsterdam, pp 83–92

Bourlet Y, Bèhar G, Guillemot F, Fréchin N, Billault A, Chaussé AM, Zoorob R, Auffray C (1988) Isolation of chicken major histocompatibility complex class II (B-L) β chain

sequences. Comparison with mammalian β chains and expression in lymphoid organs. EMBO J 7: 1031–1039
Brown JH, Jardetzky T, Saper MA, Samraoui B, Bjorkman PJ, Wiley DC (1988) A hypothetical model of the foreign antigen binding site of class II histocompatibility molecules. Nature 332: 845–850
Brown JH, Jardetzky TS, Gorga JC, Stern LJ, Urban BG, Strominger JL, Wiley DC (1993) Three-dimensional structure of the human class II histocompatibility antigen HLA-DR1. Nature 364: 33–39
Burmeister WP, Gastinel LN, Simister NE, Blum ML, Bjorkman PJ (1994) Crystal structure at 2.2 Å resolution of the MHC-related Reonatal Fc receptor. Nature 372: 336–343
Burnet FM (1971) "Self recognition" in colonial marine forms and flowering plants in relation to the evolution of immunity. Nature 232: 230–235
Calabi F, Milstein C (1986) A novel family of human major histocompatibility complex-related genes not mapping to chromosome 6. Nature 323: 540–543
Calabi F, Jarvis JM, Martin L, Milstein C (1989) Two classes of CD1 genes. Eur J Immunol 19: 285–292
Cashman NR, Pouliot Y (1990) EBV Ig-like domains. Nature 343: 319–319
Caspi RR, Avtalion RR (1984) The mixed leukocyte reaction (MLR) in carp: biodirectional and unidirectional MLR responses. Dev Comp Immunol 8: 631–637
Charles RS, Walz DA, Edwards BFP (1989) The three-dimensional structure of bovine platelet factor 4 at 3.0-Å resolution. J Biol Chem 264: 2092–2099
Cohen N (1979) Evolution of the major histocompatibility complex in vertebrates: a saga of convergent gene evolution. Transplant Proc 11: 1118–1122
Cohen N, Borysenko M (1970) Acute and chronic graft rejection: possible phylogeny of transplantation antigens. Transplant Proc 2: 333–336
Cooper EL, Rinkevich B, Uhlenbruck G, Valembois P (1992) Invertebrate immunity: another viewpoint. Scand J Immunol 35: 247–266
Cunningham BA, Hemperly JJ, Murray BA, Prediger EA, Brackenbury R, Edelman GM (1987) Neural cell adhesion molecule: structure, immunoglobulin-like domains, cell surface modulation, and alternative RNA splicing. Science 236: 799–806
De Bruijn MHL, Fey GH (1985) Human complement component C3: cDNA coding sequence and derived primary structure. Proc Natl Acad Sci USA 82: 708–712
Du Pasquier L (1993) Evolution of the immune system. In: Paul WE (ed) Fundamental immunology. Raven Press, New York, pp 199–233
Edelman GM (1988) Morphoregulatory molecules. Biochemistry 27: 3533–3543
Edmundson AB, Ely KR, Abola EE, Schiffer M, Panagiotopoulos N (1975) Rotational allomerism and divergent evolution of domains in immunoglobulin light chains. Biochemistry 14: 3953–3961
Fellah JS, Kerfourn F, Guillet F, Charlemagne J (1993) Conserved structure of amphibian T-cell antigen receptor β chain. Proc Natl Acad Sci USA 90: 6811–6814
Flajnik MF, Canel C, Kramer J, Kasahara M (1991) Which came first, MHC class I or class II? Immunogenetics 33: 295–300
Fujii T, Nakamura T, Sekizawa A, Tomonaga S (1992) Isolation and characterization of a protein from hagfish serum that is homologous to the third component of the mammalian complement system. J Immunol 148: 117–123
Garbe JC, Yang E, Fristrom JW (1993) IMP-L2: an essential secreted immunoglobulin family member implicated in neural and ectodermal development in *Drosophila*. Development 119: 1237–1250
Gardinier MV, Amiguet P, Linington C, Matthieu JM (1992) Myelin/oligodendrocyte glycoprotein is a unique member of the immunoglobulin superfamily. J Neuro Res 33: 177–187
Glazer L, Shilo BZ (1991) The *Drosophila* FGF-R homolog is expressed in the embryonic tracheal system and appears to be required for directed tracheal cell extension. Genes Dev 5: 697–705
Göbel TWF, Chen CH, Lahti J, Kubota T, Kuo CL, Aebersold R, Hood L, Cooper MD (1994) Identification of T-cell receptor α-chain genes in the chicken. Proc Natl Acad Sci USA 91: 1094–1098

Green P, Lipman D, Hillier L, Waterston R, States D, Claverie JM (1993) Ancient conserved regions in new gene sequences and the protein databases. Science 259: 1711–1716
Greenberg AS, Avila D, Hughes M, Hughes A, McKinney EC, Flajnik MF (1995) A new antigen receptor gene family that undergoes rearrangement and extensive somatic diversification in sharks. Nature 374: 168–173
Grenningloh G, Rehm EJ, Goodman CS (1991) Genetic analysis of growth cone guidance in *Drosophila*: fasciclin II functions as a neuronal recognition molecule. Cell 67: 45–57
Grossberger D, Parham P (1992) Reptilian class I major histocompatibility complex genes reveal conserved elements in class I structure. Immunogenetics 36: 166–174
Guillemot F, Billault A, Pourquié O, Béhar G, Chaussé AM, Zoorob R, Kreibich G, Auffray C (1988) A molecular map of the chicken major histocompatibility complex: the class II β genes are closely linked to the class I genes and the nucleolar organizer. EMBO J 7: 2775–2785
Hall AK, Rao MS (1992) Cytokines and neurokines: related ligands and related receptors. Trends Neurosci 15: 35–37
Hanley PJ, Seppelt IM, Gooley AA, Hook JW, Raison RL (1990) Distinct Ig H chains in a primitive vertebrate, *Eptatretus stouti.* J Immunol 145: 3823–3828
Hanley PJ, Hook JW, Raftos DA, Gooley AA, Trent R, Raison RL (1992) Hagfish humoral defense protein exhibits structural and functional homology with mammalian complement components. Proc Natl Acad Sci USA 89: 7910–7914
Harrelson AL, Goodman CS (1988) Growth cone guidance in insects: fasciclin II is a member of the immunoglobulin superfamily. Science 242: 700–708
Hashimoto K, Kurosawa Y (1991) Evolution of MHC domains: strategy for isolation of MHC genes from primitive animals. In: Klein J, Kelin D (eds) Molecular evolution of the major histocompatibility complex. NATO ASI ser H Cell Biol, vol 59. Springer, Berlin Heidelberg New York, pp 103–109
Hashimoto K, Nakanishi T, Kurosawa Y (1990) Isolation of carp genes encoding major histocompatibility complex antigens. Proc Natl Acad Sci USA 87: 6863–6867
Hashimoto K, Nakanishi T, Kurosawa Y (1992) Identification of a shark sequence resembling the major histocompatibility complex class I $\alpha 3$ domain. Proc Natl Acad Sci USA 89: 2209–2212
Hashimoto K, Hirai M, Kurosawa Y (1995) A gene outside the human MHC related to classical HLA class I genes. Science 269: 693–695
He WZ, Newell WR, Haris PI, Chapman D, Barber J (1991) Protein secondary structure of the isolated photosystem II reaction center and conformational changes studied by Fourier transform infrared spectroscopy. Biochemistry 30: 4552–4559
Hedrick SM, Nielsen EA, Kavaler J, Cohen DI, Davis MM (1984) Sequence relationships between putative T-cell receptor polypeptides and immunoglobulins. Nature 308: 153–158
Herskowitz I (1989) A regulatory hierarchy for cell specialization in yeast. Nature 342: 749–757
Hilbich C, Kisters-Woike B, Reed J, Masters CL, Beyreuther K (1991) Aggregation and secondary structure of synthetic amyloid βA4 peptides of Alzheimer's disease. J Mol Biol 218: 149–163
Hildemann WH (1970) Transplantation immunity in fishes: agnatha, chondrichthyes and osteichthyes. Transplant Proc 2: 253–259
Hildemann WH (1977) Specific immunorecognition by histocompatibility markers: the original polymorphic system of immunoreactivity characteristic of all multicellular animals. Immunogenetics 5: 193–202
Hildemann WH, Thoenes GH (1969) Immunological responses of Pacific hagfish. I. Skin transplantation immunity. Transplantation 7: 506–521
Hildemann WH, Johnson IS, Jokiel PL (1979) Immunocompetence in the lowest metazoan phylum: transplantation immunity in sponges. Science 204: 420–422
Hill RL, Delaney R, Fellows RE, Lebovitz HE (1966) The evolutionary origins of the immunoglobulins. Proc Natl Acad Sci USA 56: 1762–1769
Hinds KR, Litman GW (1986) Major reorganization of immunoglobulin V_H segmental elements during vertebrate evolution. Nature 320: 546–549

Holmgren A, Bränden CI (1989) Crystal structure of chaperone protein PapD reveals an immunoglobulin fold. Nature 342: 248–251
Holmgren A, Kuehn MJ, Bränden CI, Hultgren SJ (1992) Conserved immunoglobulin-like features in a family of periplasmic pilus chaperones in bacteria. EMBO J 11: 1617–1622
Hunkapiller T, Hood L (1989) Diversity of the immunoglobulin gene superfamily. Adv Immunol 44: 1–63
Ishiguro H, Kobayashi K, Suzuki M, Titani K, Tomonaga S, Kurosawa Y (1992) Isolation of a hagfish gene that encodes a complement component. EMBO J 11: 829–837
Jerne NK (1974) Towards a network theory of the immune system. Ann Immunol (Inst Pasteur) 125: 373–389
Jessell TM (1988) Adhesion molecules and the hierarchy of neural development. Neuron 1: 3–13
Johnson P, Williams AF (1986) Striking similarities between antigen receptor J piecies and sequence in the second chain of the mouse CD8 antigen. Nature 323: 74–76
Jones EY, Davis SJ, Williams AF, Harlos K, Stuart DI (1992) Crystal structure at 2.3 Å resolution of a soluble form of the cell adhesion molecule CD2. Nature 360: 232–239
Juy M, Amit AG, Alzari PM, Poljak RJ, Claeyssens M, Béguin P, Aubert JP (1992) Three-dimensional structure of a thermostable bacterial cellulase. Nature 357: 89–91
Kaastrup P, Nielsen B, Hoerlyck V, Simonsen M (1988) Mixed lymphocyte reactions (MLR) in rainbow trout (*Salmo gairdneri*) siblings. Dev Comp Immunol 12: 801–808
Kabat EA, Wu TT, Perry HM, Gottesman KS, Foeller C (1991) Sequences of proteins of immunological interest. Department of Health and Human Services, NIH, Bethesda, Maryland
Kang J, Lemaire HG, Unterbech A, Salbaum JM, Masters CL, Grzeschik KH, Multhaup G, Beyreuther K, Müller-Hill B (1987) The precursor of Alzheimer's disease amyloid A4 protein resembles a cell surface receptor. Nature 325: 733–736
Karlstrom RO, Wilder LP, Bastiani MJ (1993) Lachesin: an immunoglobulin superfamily protein whose expression correlates with neurogenesis in grasshopper embryos. Development 118: 509–522
Kasahara M, Vazquez M, Sato K, McKinney EC, Flajnik MF (1992) Evolution of the major histocompatibility complex: isolation of class II A cDNA clones from the cartilaginous fish. Proc Natl Acad Sci USA 89: 6688–6692
Kaufman J, Salomonsen J (1992) B-G: we know that it is, but what does it do? Immunol Today 13: 1–3
Kaufman J, Skjoedt K, Salomonsen J (1990) The MHC molecules of nonmammalian vertebrates. Immunol Rev 113: 83–117
Klämbt C, Glazer L, Shilo BZ (1992) Breathless, a *Drosophila* FGF receptor homolog, is essential for migration of tracheal and specific midline glial cells. Genes Dev 6: 1668–1678
Klein J, Satta Y, O'hUigin C, Takahata N (1993) The molecular descent of the major histocompatibility cmplex. Annu Rev Immunol 11: 269–295
Kobayashi K, Tomonaga S, Hagiwara K (1985) Isolation and characterization of immunoglobulin of hagfish, *Eptatretus burgeri*, a primitive vertebrate. Mol Immunol 22: 1091–1097
Kokubo F, Litman R, Shamblott MJ, Hinds K, Litman GW (1988) Diverse organization of immunoglobulin V_H gene loci in a primitive vertebrate. EMBO J 7: 3413–3422
Koller BH, Orr HT (1985) Cloning and complete sequence of an HLA-A2 gene: analysis of two HLA-A alleles at the nucleotide level. J Immunol 134: 2727–2733
Kolodkin AL, Matthes DJ, O'Connor TP, Patel NH, Admon A, Bentley D, Goodman CS (1992) Fasciclin IV: sequence, expression, and function during growth cone guidance in the grasshopper embryo. Neuron 9: 831–845
Kolodkin AL, Matthes DJ, Goodman CS (1993) The semaphorin genes encode a family of transmembrane and secreted growth cone guidance molecules. Cell 75: 1389–1399
Kuehn MJ, Ogg DJ, Kihlberg J, Slonim LN, Flemmer K, Bergfors T, Hultgren SJ (1993) Structural basis of pilus subunit recognition by the PapD chaperone. Science 262: 1234–1241
Lai C, Brown MA, Nave KA, Noronha AB, Quarles RH, Bloom FE, Milner RJ, Sutcliffe JG (1987) Two forms of 1B236/myelin-associated glycoprotein, a cell adhesion molecule for postnatal neural development, are produced by alternative splicing. Proc Natl Acad Sci USA 84: 4337–4341

Larhammar D, Schenning L, Gustafsson K, Witman K, Claesson L, Rask L, Peterson PA (1982) Complete amino acid sequence of an HLA-DR antigen-like β chain as predicted from the nucleotide sequence: similarities with immunoglobulins and HLA-A, -B and -C antigens. Proc Natl Acad Sci USA 79: 3687–3691
Leahy DJ, Axel R, Hendrickson WA (1992) Crystal structure of a soluble form of the human T cell corecepor CD8 at 2.6 Å, resolution Cell 68: 1145–1162
Lemke G, Axel R (1985) Isolation and sequence of a cDNA encoding the major structural protein of peripheral myelin. Cell 40: 50–1–508
Lesk AM, Chothia C (1982) Evolution of proteins formed by β-sheets II. The core of the immunoglobulin domains. J Mol Biol 160: 325–342
Leung-Hagesteijn C, Spence AM, Stern BD, Zhou Y, Su MW, Hedgecock EM, Culotti JG (1992) UNC-5, a transmembrane protein with immunoglobulin and thrombospondin type 1 domains, guides cell and pioneer axon migrations in *C. elegans*. Cell 71: 289–299
Lipke PN, Wojciechowicz D, Kurjan J (1989) AGα1 is the sructural gene for the *Saccharomyces cerevisiae* α-agglutinin, a cell surface glycoprotein involved in cell–cell interaction during mating. Mol Cell Biol 9: 3155–3165
Liszewski MK, Atkinson JP (1993) The complement system. In: Paul WE (ed) Fundamental immunology. Raven Press, New York, pp 917–939
Litman GW, Marchalonis JJ (1982) Evolution of antibodies. In: Ruben LN, Gershwin (eds) Immune regulation. Marcel Dekker, New York, pp 29–60
Litman GW, Berger L, Murphy K, Litman R, Hinds K, Erickson BW (1985) Immunoglobulin V_H gene structure and diversity in *Heterodontus*, a phylogenetically primitive shark. Proc Natl Acad Sci USA 82: 2082–2086
Malissen M, Hunkapiller T, Hood L (1983) Nucleotide sequence of a light chain gene of the mouse I-A subregion: $A\beta^d$. Science 221: 750–754
Marchalonis JJ, Edelman GM (1968) Phylogenetic origins of antibody structure III. Antibodies in the primary immune response of the sea lamprey, *Petromyzon marinus*. J Exp Med 127: 891–914
Marchalonis JJ, Schluter SF (1989) Immunoproteins in evolution. Dev Comp Immunol 13: 285–301
Miller NW, Sizemore RC, Clem LW (1985) Phylogeny of lymphocyte heterogeneity: the cellular requirements for in vitro antibody responses of channel catfish leukocytes. J Immunol 134: 2884–2888
Miller NW, Deuter A, Clem LW (1986) Phylogeny of lymphocyte heterogeneity: the cellular requirements for the mixed leukocyte reaction with channel catfish. Immunology 59: 123–128
Morgan B, Thorpe LW, Marchetti D, Perez-Polo JR (1989) Expression of nerve grow factor receptors by human peripheral blood mononuclear cells. J Neurosci Res 23: 41–45
Nakanishi T (1987) Histocompatibility analyses in tetraploids induced from clonal triploid crucian carp and in gynogenetic diploid goldfish. J Fish Biol 31: 35–40
Nonaka M, Takahashi M (1992) Complete complementary DNA sequence of the third component of complement of lamprey. Implication for the evolution of thioester containing proteins. J Immunol 148: 3290–3295
Nonaka M, Fujii T, Kaidoh T, Natsuume-Sakai S, Nonaka M, Yamaguchi N, Takahashi M (1984) Purification of a lamprey complement protein homologous to the third component of the mammalian complement system. J Immunol 133: 3242–3249
Ohno S (1970) Evolution by gene duplication. Springer Berlin Heidelberg New York
Ohno S (1990) The first set of antigens confronted by the emerging immune system. In: Kimishige I, Lachman PJ, Lerner R, Waksman BH (eds) Chemic/al immunology. Karger, Basel, pp 21–34
Oka H, Watanabe H (1957) Colony-specificity in compound ascidians as tested by fusin experiments. Proc Jpn Acad Sci 33: 657–659
Okamura K, Nakanishi T, Kurosawa Y, Hashimoto K (1993) Expansion of genes that encode MHC class I molecules in cyprinid fishes. J Immunol 151: 188–200
Overduin M, Harvey TS, Bagby S, Tong KI, Yau P, Takeichi M, Ikura M (1995) Solution structure of the epitherial cadherin domain responsible for selective cell adhesion. Science 267: 386–389

Pan KM, Baldwin M, Nguyen J, Gasset M, Serban A, Groth D, Mehlhorn I, Huang Z, Fletterick RJ, Cohen FE, Prusiner SB (1993) Conversion of α-helices into β-sheets features in the formation of the scrapie prion proteins. Proc Natl Acad Sci USA 90: 10962–10966
Patterson PH, Nawa H (1993) Neuronal differentiation factors/cytokines and synaptic plasticity. Cell (Suppl) 72: 123–137
Perey DYE, Finstad J, Pollara B, Good RA (1968) Evolution of the immune response. VI. First- and second-set skin homograft rejections in primitive fishes. Lab Invest 19: 591–597
Pham-Dinh D, Mattei MG, Nussbaum JL, Roussel G, Pontarotti P, Roeckel N, Mather IH, Artzt K, Lindahl KF, Dautigny A (1993) Myelin/oligodendrocyte glycoprotein is a member of a subset of the immunoglobulin superfamily encoded within the major histocompatibility complex. Proc Natl Acad Sci USA 90: 7990–7994
Pulido D, Campuzano S, Koda T, Modolell J, Barbacid M (1992) Dtrk, a *Drosophila* gene related to the trk family of neurotrophin receptors, encodes a novel class of neural cell adhesion molecule. EMBO J 11: 391–404
Raison RL, Hull CJ, Hildemann WH (1978) Characterization of immunoglobulin from the Pacific hagfish, a primitive vertebrate. Proc Natl Acad Sci USA 75: 5679–5682
Ramos RGP, Igloi GL, Lichte B, Baumann U, Maier D, Schneider T, Brandstätter JH, Fröhlich A, Fischbach KF (1993) The irregular chiasm C-roughest locus of *Drosophila*, which effects axonal projections and programmed cell death, encodes a novel immunoglobulin-like protein. Genes Dev 7: 2533–2547
Rast JP, Litman GW (1994) T-cell receptor gene homologs are present in the most primitive jawed vertebrates. Proc Natl Acad Sci USA 91: 9248–9252
Rippmann F, Taylor WR, Rothbard JB, Green NM (1991) A hypothetical model for the peptide binding domain of hsp70 based on the peptide binding domain of HLA. EMBO J 10: 1053–1059
Rot A (1992) Endothelial cell binding of NAP-1/IL-8: role in neutrophil emigration. Immunol Today 13: 291–294
Rudikoff S, Pumphrey JG (1986) Functional antibody lacking a variable-region disulfide bridge. Proc Natl Acad Sci USA 83: 7875–7878
Ryu SE, Kwong PD, Truneh A, Porter TG, Arthos JH, Rosenberg M, Dai X, Xuong NH, Axel R, Sweet RW, Hendrickson WA (1990) Crystal structure of an HIV-binding recombinant fragment of human CD4. Nature 348: 419–426
Sakano H, Hüppi K, Heinrich G, Tonegawa S (1979) Sequences at the somatic recombination sites of immunoglobulin light-chain genes. Nature 280: 288–294
Sakano H, Kurosawa Y, Weigert M, Tonegawa S (1981) Identification and nucleotide sequence of a diversity DNA segment (D) of immunoglobulin heavy-chain genes. Nature 290: 562–565
Scofield VL, Schlumpberger JM, West LA and Weissman IL (1982) Protochordate allorecognition is controlled by a MHC-like gene system. Nature 295: 499–502
Seeger MA, Haffley L, Kaufman TC (1988) Characterization of amalgam: a member of the immunoglobulin superfamily from *Drosophila*. Cell 55: 589–600
Seulberger H, Lottspeich F, Risau W (1990) The inducible blood-brain barrier specific molecule HT7 is a novel immunoglobulin-like cell surface glycoprotein. EMBO J 9: 2151–2158
Shishido E, Higashijima S, Emori Y, Saigo K (1993) Two FGF-receptor homologues of *Drosophila* – one is expressed in mesodermal primordium in early embryos. Development 117: 751–761
Shiue L, Gorman SD, Parnes JR (1988) A second chain of human CD8 is expressed on peripheral blood lymphocytes. J Exp Med 168: 1993–2005
Simister NE, Mostov KE (1989) An Fc receptor structurally related to MHC class I antigens. Nature 337: 184–187
Simpso TL (1984) The cell biology of sponges. Springer, Berlin Heidelberg New York
Snow PM, Bieber AJ, Goodman CS (1989) Fasciclin III: a novel homophilic adhesion molecule in *Drosophila*. Cell 59: 313–323
Sottrup-Jensen L, Stepanik TM, Kristensen T, Lønblad PB, Jones CM, Wierzbicki DM, Magnusson S, Domdey H, Wetsel RA, Lundwall A, Tack BF, Fey GH (1985) Common evolutionary origin of α2-macroglobulin and complement components C3 and C4. Proc Natl Acad Sci USA 82: 9–13

Springer TA (1994) Traffic signals for lymphocyte recirculation and leukocyte emigration: the multistep paradigm. Cell 76: 301–314
Srivastava PK, Udono H, Blachere NE, Li Z (1994) Heat shock proteins transfer peptides during antigen processing and CTL priming. Immunogenetics 39: 93–98
Stein JC, Howlett B, Boyes DC, Nasrallah ME, Nasrallah JB (1991) Molecular cloning of a putative receptor protein kinase gene encoded at the self-incompatibility locus of *Brassica oleracea*. Proc Natl Acad Sci USA 88: 8816–8820
Steinman L (1993) Connections between the immune system and the nervous system. Proc Natl Acad Sci USA 90: 7912–7914
Sun SC, Lindström I, Boman HG, Faye I, Schmidt O (1990) Hemolin: an insect-immune protein belonging to the immunoglobulin superfamily. Science 250: 1729–1732
Takeichi M (1990) Cadherin cell adhesion receptors as a morphogenetic regulator. Science 251: 1451–1455
Tjoelker LW, Carlson LM, Lee K, Lahti J, McCormack WT, Leiden JM, Chen CH, Cooper MD, Thompson CB (1990) Evolutionary conservation of antigen recognition. The chicken T-cell receptor β chain. Proc Natl Acad Sci USA 87: 7856–7860
Tonegawa S (1983) Somatic generation of antibody diversity. Nature 302: 575–581
Valegård K, Liljus L, Fridborg K, Unge T (1990) The three-dimensional structure of the bacterial virus MS2. Nature 345: 36–41
Varner J, Neame P, Litman GW (1991) A serum heterodimer from hagfish (*Eptatretus stoutii*) exhibits structural similarity and partial sequence identity with immunoglobulin. Proc Natl Acad Sci USA 88: 1746–1750
Wang H, Wu JJ, Tang P (1989) Superfamily expands. Nature 337: 514–514
Wang J, Yan Y, Garrett TPJ, Liu J, Rodgers DW, Garlick RL, Tarr GE, Husain Y, Reinherz EL, Harrison SC (1990) Atomic structure of a fragment of human CD4 containing two immunoglobulin-like domains. Nature 348: 411–418
Weiss E, Golden L, Zakut R, Mellor A, Fahrner K, Kvist S, Flavell RA (1983) The DNA sequence of the H-2K^b gene: evidence for gene conversion as a mechanism for the generation of polymorphism in histocompatibility antigens. EMBO J 2: 453–462
Weissman IL, Saito Y, Rinkevich B (1990) Allorecognition histocompatibility in protochordate species: is the relationship to MHC somatic or structural? Immunol Rev 113: 227–241
Wetsel RA, Ogata RT, Tack BF (1987) Primary structure of the fifth component of murine complement. Biochemistry 26: 737–743
Williams AF (1982) Surface molecules and cell interactions. J Theor Biol 98: 221–234
Williams AF (1984) Molecules in the immunoglobulin superfamily. Immunol Today 5: 219–221
Williams AF (1987) A year in the life of the immunoglobulin superfamily. Immunol Today 8: 298–303
Williams AF, Barclay AN (1988) The immunoglobulin superfamily domains for cell surface recognition. Annu Rev Immunol 6: 381–405
Williams AF, Barclay AN (1989) The immunoglobulin superfamily. In: Honjo T, Alt FW, Rabitts TH (eds) Immunoglobulin genes. Academic Press, London, pp 361–387
Williams AF, Gagnon J (1982) Neuronal cell Thy-1 glycoprotein: homology with immunoglobulin. Science 216: 696–703
Wojciechowicz D, Lu CF, Kurjan J, Lipke PN (1993) Cell surface anchorage and ligand-binding domains of the *Saccharomyces cerevisiae* cell adhesion protein α-agglutinin, a member of the immunoglobulin superfamily. Mol Cell Biol 13: 2554–2563
Yanagi Y, Yoshikai Y, Leggett K, Clark SP, Aleksander I, Mak TW (1984) A human T cell-specific cDNA clone encodes a protein having extensive homology to immunoglobulin chains. Nature 308: 145–149
Yayon A, Zimmer Y, Guo-Hong S, Avivi A, Yarden Y, Givol D (1992) A confined variable region confers ligand specificity on fibroblast growth factor receptors: implications for the origin of the immunoglobulin fold. EMBO J 1: 1885–1890

Chapter 6

Insect Hemolymph Proteins from the Ig Superfamily

M.R. Kanost and L. Zhao

Contents

1 Introduction

Antibodies have not been identified in invertebrates. It appears that rearranging immunoglobulin (Ig) genes and clonal selection arose in vertebrates after their evolutionary divergence from the invertebrates (Marchalonis and Schluter 1990). However, insects have been shown to mount a humoral response to bacterial infections, which includes the synthesis and secretion of a battery of antibacterial proteins (Dunn 1990; Hultmark 1993). It has recently been recognized through analysis of amino acid sequences deduced from cloned cDNAs that one of the hemolymph proteins induced by bacteria in the lepidopteran insects *Hyalophora cecropia* and *Manduca sexta* is a member of the Ig superfamily (Sun et al. 1990; Ladendorff and Kanost 1991). This protein, which has been named hemolin (previously known as P4), has no direct antibacterial activity but may function in recognition of bacteria and/or in modulation of the adhesive properties of hemocytes. Hemolin is composed of four Ig domains (see Sect. 4) which have more similarity in amino acid sequence to Ig domains from cell adhesion molecules than to those from antibodies. We will review the short history of research on hemolin and pose some hypotheses regarding the functions of hemolin in insect immunity.

Department of Biochemistry, Kansas State University, Manhatan, Kansas 66506, USA

Advances in Comparative and Environmental Physiology, Vol. 23

2 Properties of Hemolin

Hemolin was first identified as a protein induced by bacteria in diapausing pupae of *H. cecropia* (Faye et al. 1975) and designated protein P4 (the fourth induced protein band on an SDS polyacrylamide gel, counting from high to low molecular weight). Hemolin, isolated from hemolymph of bacteria-injected, diapausing pupae of *H. cecropia* or larvae of *M. sexta* has a $M_r = 48\,000$ and $pI = 8.2$–8.4 (Rasmuson and Boman 1979; Ladendorff and Kanost 1990). Hemolin exists in more than one isoform with slightly different isoelectric points in both *H. cecropia* (Andersson and Steiner 1987) and *M. sexta* (M11 proteins in Hughes et al. 1983; Hurlbert et al. 1985). Whether these represent different alleles or products of distinct genes is not yet clear. Hemolin from *M. sexta* is a glycoprotein, containing 12% carbohydrate and binds to the lectin concanavalin A (Ladendorff and Kanost 1990); its amino acid sequence contains one N-linked glycosylation site (Ladendorff and Kanost 1991). Carbohydrate has not been detected in hemolin from *H. cecropia* (Andersson and Steiner 1987), although two potential N-linked glycosylation sites are present in its amino acid sequence (Sun et al. 1990; I. Faye, per. comm.). Analysis by circular dichroism spectroscopy, indicating that hemolins from both *H. cecropia* and *M. sexta* are composed predominantly of antiparallel β-sheets (Andersson and Steiner 1987; Kanost et al. 1994), is compatible with the structures known for members of the Ig superfamily (Hunkapiller and Hood 1989).

A protein related to hemolin has been isolated from untreated *M. sexta* adults and named postlarval protein (Ryan et al. 1988). Postlarval protein has a $M_r = 50\,000$ and $pI = 8.6$ and is not detectable in hemolymph until after the end of the feeding stage of the last larval instar. It increases in concentration during the prepupal stage and is present in hemolymph throughout the pupal and adult stages. The amino acid composition of postlarval protein is very similar to that of hemolin, and antibodies to hemolin cross-react with postlarval protein and vice versa (Ladendorff and Kanost 1990). Postlarval protein differs from hemolin in some chemical properties in addition to its distinct pattern of expression. Postlarval protein exhibits a slightly slower migration than hemolin in SDS polyacrylamide gel electrophoresis, postlarval protein does not contain detectable carbohydrate and does not bind to concanavalin A, and postlarval protein is blocked at its NH_2 terminus, whereas hemolin is susceptible to Edman degradation (Ladendorff and Kanost 1990; X. Yu and M.R. Kanost, unpubl. results).

3 Regulation of Hemolin Expression

In larvae of *M. sexta* and diapausing pupae of *H. cecropia*, hemolin is present at a low, constitutive level in uninfected (naive) insects. The concentration of hemolin in hemolymph of naive *H. cecropia* is approximately 0.38 mg/ml and after injection of *Enterobacter cloacae* increases 18-fold over a period of 10 days, reaching 7 mg/ml (Andersson and Steiner 1987). In naive *M. sexta* larvae, hemolin is present at 35 μg/ml, tenfold lower in concentration in comparison with *H. cecropia* pupae. After injection of *M. sexta* with either Gram-positive or Gram-negative bacteria, hemolin concentration increases to a maximum of 1.1–1.5 mg/ml in 2 days (Ladendorff and Kanost 1990). Both peptidoglycan and lipopolysaccharide (LPS), isolated components of bacterial cell envelopes, also elicit increased levels of hemolin in *M. sexta*, as does zymosan, a preparation of yeast cell walls (Ladendorff and Kanost 1990).

The primary tissue source of hemolin is the fat body. Pupal fat body from *H. cecropia* was shown to synthesize hemolin in tissue culture, and addition of heat-killed bacteria or LPS to the culture results in increased levels of hemolin synthesis (Trenczek and Faye 1988). Addition of hemocytes from insects vaccinated with LPS to the fat body cultures also stimulates hemolin synthesis, indicating a potential for communication between hemocytes and fat body in regulation of hemolin synthesis. Hemocytes themselves also synthesize hemolin after exposure to bacteria, both in *H. cecropia* (Trenczek 1988) and in *M. sexta* (Wang et al. 1995). In *M. sexta*, hemolin mRNA can be detected by Northern blot analysis of hemocyte RNA isolated after injection of larvae with bacteria, and immunohistochemistry demonstrates that hemolin accumulates in granules of granulocytes after injection of bacteria.

In *M. sexta*, hemolin mRNA is not detectable by Northern hybridization in fat body of naive larvae, but is observed by 1 h after injection of bacteria. Hemolin mRNA increases in abundance for up to 24 h after injection, becoming a very abundant mRNA in fat body. Hemolin mRNA also becomes elevated in fat body of adult *M. sexta* after injection of bacteria, with the following differences from larvae: an RNA hybridizing to hemolin DNA probes is present in fat body from naive adults, and increased levels of hemolin mRNA are not apparent until four hours after injection of bacteria (Ladendorff and Kanost 1991).

Much remains to be learned about the molecular mechanisms of regulation of hemolin genes and their induced transcription by bacterial cell surface components. Several other bacteria-induced genes for insect hemolymph proteins contain an upstream sequence similar to the binding site for the mammalian transcription factor NF-κB (Hultmark 1993), and a nuclear factor that specifically binds to these sites has been isolated from *H. cecropia* fat body (Sun and Faye 1992). It seems likely that κb-like elements will also be found as potential regulatory sequences for hemolin expression. Further questions in need of investigation involve the steps preceding induced transcription. What are the recognition molecules that detect bacterial cell wall components? How is the detection of bacteria in hemolymph transmitted as a signal to fat body cells for activation of transcription factors? Is there communication between hemocytes and fat body in this process? Is hemolin involved as a recognition molecule with a role in the transfer of signals to responsive cells?

4 Sequence Analysis of Hemolin

The amino acid sequences deduced from hemolin cDNAs include an NH_2-terminal signal peptide of 17–18 residues followed by the mature protein containing 394–395 residues. Within the mature hemolin protein are four repeated units of 90–100 residues that have the sequence characteristics of Ig domains (Sun et al. 1990; Ladendorff and Kanost 1991), which define members of the Ig superfamily. Ig domains (also called Ig homology units) are sequences of 70–110 residues with a highly conserved disulfide bridge formed by cysteines separated by 50–70 residues, and with a number of other relatively conserved residues that contribute to the structure known as an Ig fold (Williams and Barclay, 1988; Hunkapiller and Hood 1989; see also Y. Kurasawa, Vol. 1 of this work). Antibodies contain two types of Ig domains: C (constant) and V (variable) domains, which are grouped separately due to differences in sequence and length, but both have the conserved characters of Ig domains. A third type of Ig domain (H or C2) has been identified and has properties shared by both V and C domains. H domains are thought to represent more closely than C or V the primordial Ig motif (Hunkapiller and Hood 1989). The four Ig domains in hemolin are characteristic of the H-type unit. H-type Ig domains have been found in a number of cell adhesion molecules that are important in nervous system development, in liver cell adhesion molecules, and in human carcinoembryonic antigen and α1B-glycoprotein, whose functions are not yet known (Williams and Barclay 1988; Hunkapiller and Hood 1989). Nerve cell adhesion molecules (NCAM) composed of multiple H-type Ig domains have been identified in several vertebrates and in insects, particularly *Drosophila melanogaster*. In computer searches of protein sequence databases the sequences found to be most similar to hemolin include neuroglian from *Drosophila* (Bieber et al., 1989), neurofascin from chicken (Volkmer et al. 1992), and Nr-CAM from chicken (Grumet et al. 1991), with high similarity scores with a large number of other cell adhesion molecules from mammals, birds, *Xenopus*, and *Drosophila*. An alignment of the sequence of *M. sexta* hemolin with the first four Ig domains of a *Drosophila* protein, neuroglian (Bieber et al. 1989), a human protein, axonin (Hasler et al. 1993) and a mouse protein, NCAM (Barthels et al. 1987) is shown in Fig. 1. Although the percentage identity is low (in the range of 22–36% for each pair of sequences) statistical analysis of the similarity shows significant homology of hemolin to other members of the Ig superfamily (Ladendorff and Kanost 1991). The alignment demonstrates that the regions around the invariant cysteines are the most highly conserved and that the sequences between these regions, including their lengths, are more variable. The conserved residues near the cysteines are characteristic of the H-type Ig sequence motif described by Hunkapiller and Hood (1989).

The sequences of hemolin from *M. sexta* and *H. cecropia* are 62% identical, with a large number of conservative substitutions (Fig. 2). Domain 3 is the most conserved (77% identical) of the four Ig domains when the individual domains are compared, suggesting that this part of the protein may be important in its function. Nearly all of the members of the Ig superfamily with known functions

Ig Domain 1

```
                                        ∇            *   *                               * * * ∇    *
HEMOLIN      QPVEKMPVLKDQPAEVLFRESQATVLECVTENGDKDVKYSW QKDGKEFKWQ  EHNIAQRKDEGSLVFLKP EAKDEGKYRCFAESAAGVATSHIISFRRTYM
NEUROGLIAN  ITKQPAPGELLFKVAQQNKESDNPFIIEC EADGQPEPEYSW IKNGKKFDWQAYDNRMLRQPGRGTLVITIP KDEDRGHYQCFASNEFGTATSNSVYVRKAEL
AXONIN      TTFGPVFEDQPLSVLFPEESTEEQVLLAC RARASPPATYRW KMNGTEMKL   EPGSRHQLVGGNLVIMNPTKAQDAGVYQCLASNPVGTVVSREAILRFGFL
NCAM               LQVDIVPSQGEISVGESKFFLCQVAGDAKDKDISWFSPNGEKLSPNQQRISVVWNDDDSSTLTIYNANIDDAGIYKCVVTAEDGTQSEATVNVKIFQK
```

Ig Domain 2

```
              **      ∇       *    *                                  *         *      *∇
HEMOLIN     VVPTTFKTVEKKPVEGSWLKLECSIPEGYPKPTIVWRKQLGEDESIADSILARRITQSPE        GDLYFTSVEKEDVSESYKYVCAAKSPA IDGDVPLVGYTIKSLEKNT
NEUROGLIAN  NAFKDEAAKTLEAVEGEPFMLKCAAPDGFPSPTVNWMIQESIDGSI KSINNSRMTLDPEVKQMGVSAGNLWFSNVTREDASSDFYYACSA TSV FRSE     YKIGNKVLLD
AXONIN      QEFSKEERDPVKAHEGWGVMLPCNPPAHYPGLSYRWLLN EFPNFI   PTDGRHFVSQTT       GNLYIARTNASDLGN   YSCLATSHMDFSTKSVFSKFAQLNLAAED
NCAM        LMFKN APTPQEFKEGEDAVIVCDVVSSLP PTIIWKHK GRDVIL   KKDVRFIVLSN        NYLQIRGIKKTDEGT   YRCEGRILA  RGE    INFKDIQVIVNV
```

Ig Domain 3

```
                              *  *      ∇   * *     * *                         *      *     ∇   *  *         *   *
HEMOLIN     NQKNGELVPMYVS NDMIAKAGDVTMIYCMYGGVPMAYPNWFKDGKDVNGKPSDRITRHNRTSGKRLLIKETLLEDRGTFTCDVNNEVGKPQKHSVKLTVVSGPR
NEUROGLIAN  SQNKHPPVRQYVSRRQSLALRGKRMELFCIYGGTPLPQTVWSKDGQRI  QWSDRITQGH  YGKSLVIRQTNFDDAGTYTCDVSNGVGNAQSFSIILNVNSVPY
AXONIN      TRLFAPSIKARFP AETYALVGQQVTLECFAFGNPVPRIKWRKVDGSL   SPQWTT     AEPTLQIPSVSFEDEGTYECEAENSKGR DTVQGRIIVQAQPE
NCAM            PPTVQARQSIVNATANLGQSVTLVCDADGFPEPTMSWTKDGEPIENEEEDERSRSSVSDSSEVTIRNVDKNDEAEYVCIAENKAGE QDASIHLKVFAKPK
```

Ig Domain 4

```
                               ∇     *     *                   *                * * ∇ * * *
HEMOLIN     FTKKPEKQVIAKQARLVIPCEVSRYPAAPVSWTFNAKPI   SG      SRVVASPSGLTIKGIQKSDKGYYGCQAHNEHGDAYAETLVIVA
NEUROGLIAN  FTKEPEIATAAEDEEVVFECRAAGVPEPKISWIHNGKPIEQSTPN     PRRTVTDNTIRIINLVKGDTGNYGCNATNSLGYVYKDVYLNVQ
AXONIN      WLKVISDTEADIGSNLRWGCAAAGKPRPTVRWLRNGEPL  ASQ      NRVEVLAGDLRFSKLSLEDSGMYQCVAENKHGTIYASAELAVQ
NCAM        ITYVENQTAMELEEQVTLTCEASGDPIPSITWRTSTRNISSEEQDLDGHMVVRSHARVSSLTLKSIQYRDAGEYMCTASNTIGQDSQSIDLEFQ
```

Fig. 1. Alignment of the amino acid sequence of *M. sexta* hemolin with other members of the Ig superfamily. The first four Ig domains of *Drosophila melanogaster* neuroglian, human axonin, and mouse NCAM (all cell adhesion molecules involved in nervous system development) were aligned with the four Ig domains of hemolin, using the multiple alignment programs of Feng and Doolittle (1990). Cysteines which define the Ig domains are marked with ∇. Positions that are identical in all four sequences are marked with*

```
                                           ▽
MANDUCA       1 QPVEKMPVLKDQPAEVLFRESQATVLECVTENGDKDVKYSWQKDGKEFKWQEHNIAQRKD  60
                 ||:| |||||||||||||| ::|||||: |  |  ||||| |||| : ||||| | |||
CECROPIA      1 LPVDKYPVLKDQPAEVLFRENNPTVLECIIEGNDQGVKYSWKKDGKSYNWQEHNAALRKD  60

                                  ▽
MANDUCA      61 EGSLVFLKPEAKDEGKYRCFAESAAGVATSHIISFRRTYMVVPTTFKTVEKKPVEGSWLK 120
                |||||||:|:| ||| | |||| :|||| |::||||:||::     || || |:||  :
CECROPIA     61 EGSLVFLRPQASDEGHYQCFAETPAGVASSRVISFRKTYLIASPA.KTHEKTPIEGRPFQ 119

                  ▽
MANDUCA     121 LECSIPEGYPKPTIVWRKQLGEDESIAD.SILARRITQSPEGDLYFTSVEKEDVSESYKY 179
                |:| :|::|||| | |:| |:: :  ||   : || |  |:|:|||| | |||||: |||
CECROPIA    120 LDCVLPNAYPKPLITWKKRLSGADPNADVTDFDRRNTAGPDGNLYFTIVTKEDVSDIYKY 179

                 ▽                                                    ▽
MANDUCA     180 VCAAKSPAIDGDVPLVGYTIKSLEKNTNQKNGELVPMYVSNDMIAKAGDVTMIYCMYGGV 239
                || || :|:|::| ||:| ||:: |:     || || ||| ||:||||||||||||||:
CECROPIA    180 VCTAKNAAVDEEVVLVEYEIKGVTKDNSGYKGEPVPQYVSKDMMAKAGDVTMIYCMYGSN 239

                                                               ▽
MANDUCA     240 PMAYPNWFKDGKDVNGKPSDRITRHNRTSGKRLLIKETLLEDRGTFTCDVNNEVGKPQKH 299
                ||:|||:||:|||||| | |||||||||||||||:| || || | :||:|:|:|||||||
CECROPIA    240 PMGYPNYFKNGKDVNGNPEDRITRHNRTSGKRLLFKTTLPEDEGVYTCEVDNGVGKPQKH 299

                                               ▽
MANDUCA     300 SVKLTVVSGPRFTKKPEKQVIAKQAR.LVIPCEVSRYPAAPVSWTFNAKPISGSRVVASP 358
                |:||||||:|::  |||| :: ||:  : ||| |   ||: | |  ||||:||:|
CECROPIA    300 SLKLTVVSAPKYEQKPEKVIVVKQGQDVTIPCKVTGLPAPNVVWSHNAKPLSGGRATVTD 359

                                 ▽
MANDUCA     359 SGLTIKGIQKSDKGYYGCQAHNEHGDAYAETLVIVA 394
                ||| ||::  :||||||| | ||||| | |||| |
CECROPIA    360 SGLVIKSVKNGDKGYYGCRATNEHGDKYFETLVQVN 395
```

Fig. 2. Alignment of the amino acid sequences of hemolin from *M. sexta* and *H. cecropia*. Identical amino acids are indicated by |, and conserved substitutions are indicated by : Cysteines that mark the Ig domains are labeled with ∇. Sequences were obtained from the GenBank database (*M. sexta* hemolin accession number, M64346; *H. cecropia* hemolin accession number, M63398)

act in some type of recognition at a cell surface and bind a specific ligand. The Ig domains provide a stable framework for the binding sites, which are often located in the loops between β strands. NCAM proteins are homophilic cell adhesion molecules—they bind specifically to other NCAM molecules on a different cell. The similarity of hemolin with the NCAMs suggests that hemolin might also have a role in cell adhesion (discussed further below).

5 Possible Functions of Hemolin

The biological function of hemolin is still unclear. Hemolin has no direct antibacterial activity, and its membership in the Ig superfamily suggests instead a potential for a role in some kind of cellular recognition. Common properties of proteins in the Ig superfamily include binding to a specific ligand (which for cell adhesion molecules is usually another member of the Ig superfamily) and a role in recognition at a cell surface. Some data are beginning to indicate that hemolin may function in recognition of bacteria by hemocytes and in modulation of the adhesive properties of hemocytes, both of which should involve specific binding and action at hemocyte surfaces.

Hemolin from *H. cecropia* has been found to associate with the surface of *Escherichia coli* cells, perhaps as a complex with other hemolymph proteins (Sun et al. 1990). Binding of hemolin to *E. coli* does not depend on the presence of the carbohydrate portion of LPS and does not require divalent cations (Schmidt et al. 1993). Hemolin from *M. sexta* also associates with the surface of *E. coli* cells (Ladendorff and Kanost 1991). Although the complex described in *H. cecropia* was not observed in *M. sexta*, the lipid transfer protein lipophorin was identified as a protein from *M. sexta* hemolymph that also associates with bacterial cells. The binding of hemolin to bacteria requires much more thorough biochemical analysis, before we can understand its significance. Questions that need to be answered are: To what molecule on a bacterial cell does hemolin bind? How tight is the binding? Is the binding specific?

Hemolin has also been shown to associate with the surface of hemocytes. Immunofluorescence staining of tissues from *H. cecropia* pupae after injection of bacteria indicated that hemolin was associated with granulocytes, spherulocytes, and plasmatocytes, although the intensity of staining in plasmatocytes was variable (Andersson and Steiner 1987). Hemolin was also located in nodules formed by hemocytes in pupae injected with bacteria. *M. sexta* larval hemocytes were shown to bind hemolin in an experiment in which hemocytes from naive larvae were incubated with purified hemolin (Ladendorff and Kanost 1991). The binding of hemolin to *M. sexta* hemocytes is fairly uniform, with no apparent bias toward any cell type. When hemolin is labeled with the fluorescent marker fluorescein and incubated with hemocytes, followed by washing of the hemocytes and measurement of fluorescence by flow cytometry, the great majority of hemocytes are labeled (Fig. 3). Immunofluorescence microscopy confirms that all

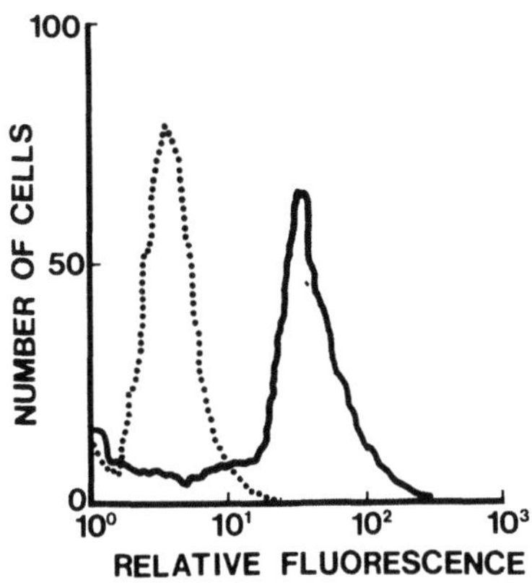

Fig. 3. Binding of hemolin to hemocytes from *M. sexta*. Hemocytes from fifth instar *M. sexta* larvae were washed with anticoagulant saline and then incubated with fluorescently labeled hemolin (FITC-hemolin, 0.1 μg/ml) for 30 min at 4 °C. Hemocytes were then washed twice with saline and fluorescence was measured by flow cytometry. The *dotted line* indicates the background fluorescence of hemocytes that were not incubated with FITC hemolin. The *solid line* indicates the fluorescence due to binding of FITC hemolin. Each plot represents analysis of 10 000 cells. Note that the entire peak shifts to higher fluorescence after incubation with hemolin, indicating that hemolin binding does not depend on morphological hemocyte type.

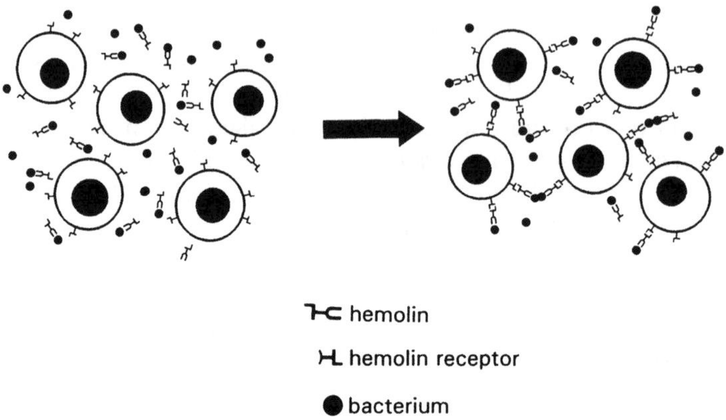

Fig. 4. A hypothetical model for the function of hemolin as an opsonin. Hemolin molecules in the hemolymph bind to the surface of bacterial cells via a binding site on the hemolin protein. Hemolin then binds (via a separate binding site) to a specific receptor on the surface of hemocytes, bringing the bacteria into close association with the hemocyte surface, facilitating phagocytosis

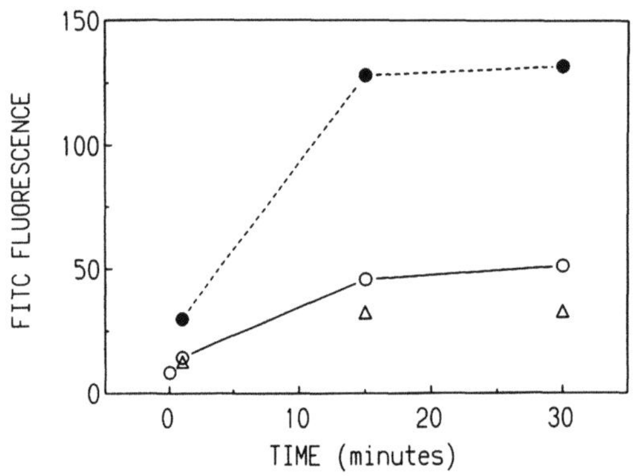

Fig. 5. Increased association of bacteria with hemocytes in the presence of hemolin. Hemocytes collected from fifth instar *M. sexta* larvae were washed to remove plasma proteins and then incubated in insect cell culture medium with FITC-labeled *E. coli* cells. At various times of incubation, association of fluorescent bacteria with the hemocytes was measured by flow cytometry. Increased fluorescence indicates higher numbers of fluorescent bacteria associated with hemocytes (the assay did not discriminate between attachment and internalization). *Open circles* are measurements in culture medium alone. *Filled circles* are measurements in the presence of hemolin (100 μg/ml). *Triangles* are measurements in the presence of ovalbumin as a control protein (100 μg/ml). Each *point* is the mean fluorescence of 10000 cells

of the morphological hemocyte types appear to contain binding sites for hemolin. However, the binding of hemolin to hemocyte surfaces also needs more investigation, particularly the identification of hemocyte surface molecules that may be involved in association with hemolin.

The observations that hemolin can associate with both bacteria and hemocytes suggests a possible role for hemolin as an opsonin, bringing bacteria to the surface of hemocytes and increasing the efficiency of phagocytosis. A model for an opsonic function of hemolin is diagramed in Fig. 4. In this model, hemolin must have two binding sites, one for bacteria and another for a receptor on the surface of hemocytes. Hemolin present in the hemolymph binds to the surface of bacteria and then the complex of bacterium–hemolin binds to a hemolin receptor, which facilitates phagocytosis of the bacteria. Preliminary experiments indicate that the presence of hemolin can increase the association of *E. coli* cells with *M. sexta* hemocytes (Fig. 5). Perhaps the low concentration of hemolin in naive insects can function in this manner early in a cellular response to bacterial infection. The continued presence of significant numbers of bacteria in the

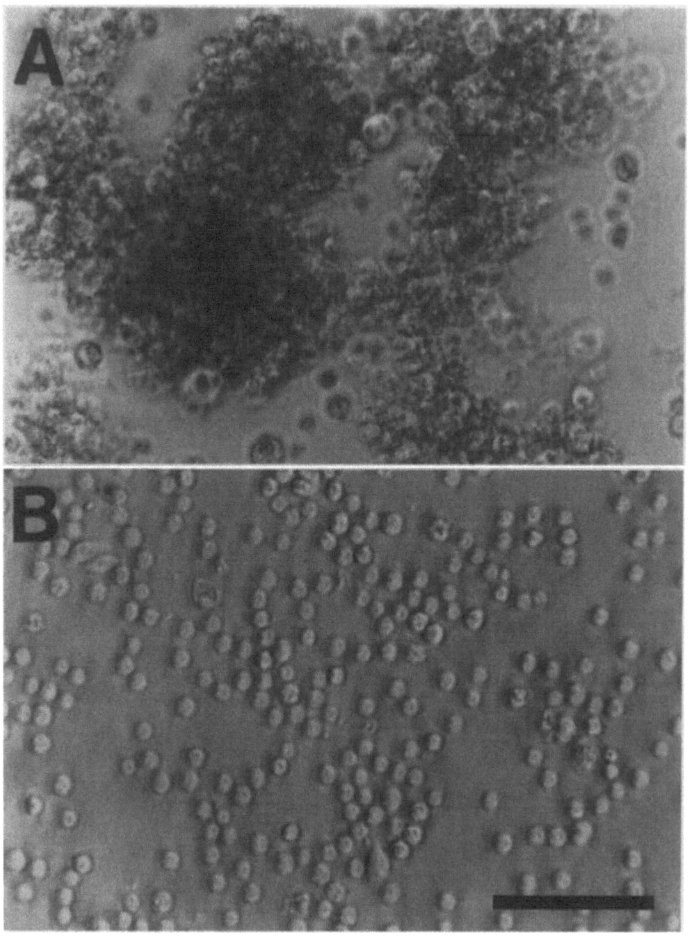

Fig. 6. Inhibition of hemocyte aggregation by hemolin. Hemocytes from a fifth instar *M. sexta* larva were incubated in wells of a 96 well plate in **A** insect cell culture medium or **B** medium containing 160 μg/ml hemolin at 24 °C for 3 h with shaking at 120 rpm. Scale *bar* = 100 μm

hemocoel could then stimulate the synthesis of hemolin by the fat body for an enhanced response to further infection.

Another biological activity that has been observed for hemolin is the inhibition of hemocyte aggregation (Ladendorff and Kanost 1991; Kanost et al. 1994). When hemocytes from *M. sexta* larvae are cultured in polystyrene plates, the cells begin to adhere to one another, and eventually form large aggregates, similar in appearance to nodules formed during bacterial infection (Fig. 6A). When hemolin is present in the medium at concentrations that are achieved during bacterial challenge, this aggregation is blocked (Fig. 6B). Hemocytes incubated in the presence of control proteins such as ovalbumin or bovine serum albumin aggregate in the same manner as hemocytes in culture medium alone, indicating that the aggregation inhibiting activity of hemolin is not due to the presence of protein in the cell culture. It is known that cell adhesion molecules modulate the adhesiveness of vertebrate leukocytes, regulating the migration and interaction of cells during an immune response (Springer 1990). Perhaps hemolin functions in regulation of the adhesiveness of insect hemocytes, either to each other or to basement membrane, in an analogous manner. A hypothetical model for the aggregation inhibitory effect of hemolin is shown in Fig. 7. During an infection or in response to a foreign surface, a wound, or other trauma, hemocytes

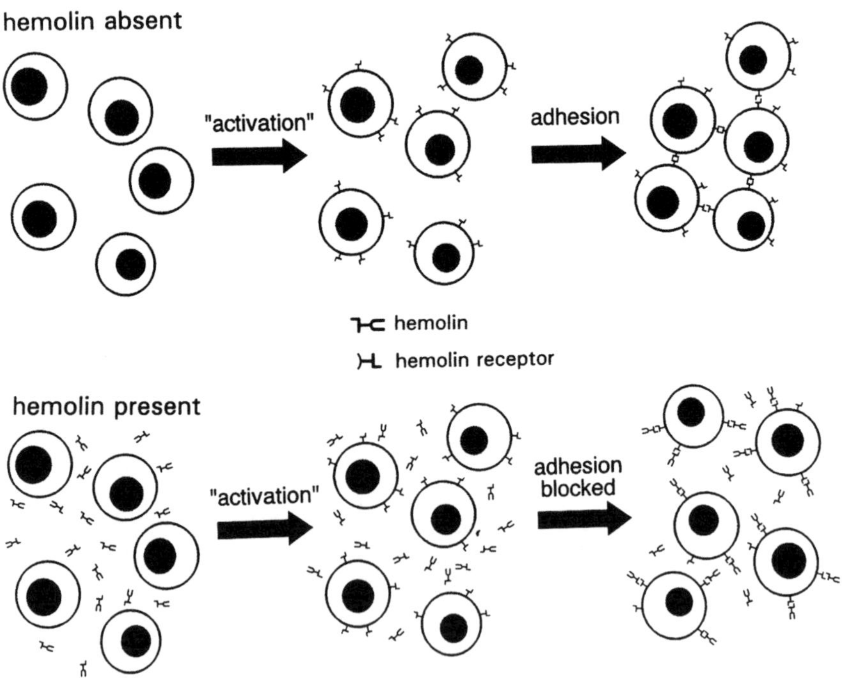

Fig. 7. A hypothetical model for the inhibition of hemocyte aggregation by hemolin. In the presence of hemolin, homophilic adhesion receptors on the surface of "activated" hemocytes are blocked, preventing aggregation of hemocytes

might become "activated," meaning that they begin to express on their surface a homophilic cell adhesion molecule. Like the vertebrate lymphocyte adhesion molecules, this might be a rapid, qualitative change to activate a receptor, rather than an increase in the number of receptor molecules. Activation of the cell adhesion molecules would result in adhesion of cells through interaction of the homophilic receptors. The model predicts that hemolin binds to the cell surface adhesion molecules, blocking the binding sites and preventing interaction between receptors on different cells. The validity of this model remains to be tested through experiments aimed at understanding the molecular mechanisms of adhesion of insect hemocytes.

6 Conclusion

The Ig gene superfamily seems to have risen early in evolution of the Animal Kingdom, encoding a very versatile type of protein. The stable framework of the

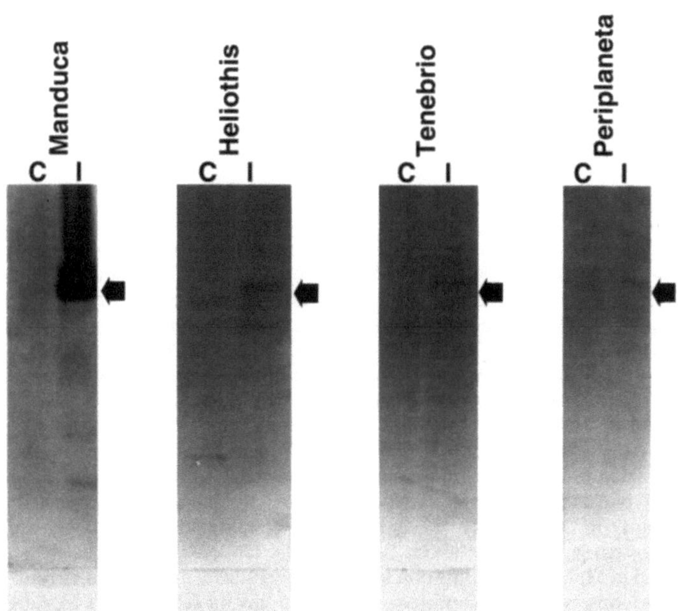

Fig. 8. Cross-reaction of *M. sexta* hemolin antiserum with hemolymph proteins from other insect species. Hemolymph samples were collected from naive insects (*lanes labeled* C) and from insects 24 h after injection of 100 μg dried *Micrococcus lysodeikticus* cell (*lanes marked* I). Samples of cell-free hemolymph were separated by SDS polyacrylamide gel electrophoresis and transferred to nitrocellulose membrane followed by detection with hemolin antiserum (rabbit) and enzyme-labeled second antibody (goat anti-rabbit IgG-alkaline phosphatase). Insects used in the experiment were *M. sexta* (order Lepidoptera) fifth instar larva, *Heliothis zea* (order Lepidoptera) fifth instar larva, *Tenebrio molitor* (order Coleoptera) last instar larva, *Periplaneta americana* (order Orthoptera) adult. *Arrows* indicate bands, of approximately the same size as hemolin, which were detected by the hemolin antiserum

Ig fold, combined with its capacity for forming specific binding sites has been selected for numerous functions involving cell–cell interactions (Hunkapiller et al. 1989). Gene duplications, followed by divergence and selection for new functions have led to the emergence of Ig gene families and an array of functions centered in the nervous and immune systems. Like vertebrates, insects have some Ig superfamily proteins which function in the nervous system and others (hemolin is the first, but may be not the last example) which are part of the immune system. Although hemolin has been studied so far only in two lepidopteran species, a preliminary experiment indicates that a bacteria-induced protein that cross-reacts with *M. sexta* hemolin antibody is present in insects from other orders (Fig. 8). As we learn more about the immune systems of invertebrates at a molecular level, it will not be surprising if other proteins composed of Ig domains are discovered to fill a broad range of roles in intercellular recognition and response to microbial infection.

Acknowledgement. Research from our laboratory was supported by NIH grant AI31084. Contribution 95-546-B from the Kansas Agricultural Experiment Station.

References

Andersson K, Steiner H (1987) Structure and properties of protein P4, the major bacteria-inducible protein in pupae of *Hyalophora cecropia*. Insect Biochem 17: 133–140

Barthels D, Santoni MJ, Wille W, Ruppert C, Chaix JC, Hirsch MR, Fontecilla-Camps JC, Goridis C (1987) Isolation and nucleotide sequence of mouse NCAM cDNA that codes for a Mr 79,000 polypeptide without a membrane-spanning region. EMBO J 6: 907–914

Bieber AJ, Snow PM, Hortsch M, Patel NH, Jacobs JR, Traquina ZR, Schilling J, Goodman CS (1989) *Drosophila* neuroglian: a member of the immunoglobulin superfamily with extensive homology to the vertebrate neural adhesion molecule L1. Cell 59: 447–460

Dunn PE (1990) Humoral immunity in insects. BioScience 40: 738–744

Faye I, Pye A, Rasmuson T, Boman HG, Boman IA (1975) Insect immunity II. Simultaneous induction of antibacterial activity and selective synthesis of some hemolymph proteins in diapausing pupae of *Hyalophora cecropia* and *Samia cynthia*. Infect Immun 12: 1426–1438

Feng D-F, Doolittle RF (1990) Progressive alignment and phylogenetic tree construction of protein sequences. Methods Enzymol 183: 375–387

Grumet M, Mauro V, Burgoon MP, Edelman GM, Cunningham BA (1991) Structure of a new nervous system glycoprotein, Nr-CAM, and its relationship to subgroups of neural cell adhesion molecules. J Cell Biol 113: 1399–1412

Hasler TH, Rader C, Stoeckli ET, Zuellig RA, Sonderegger P (1993) cDNA cloning, structural features, and eucaryotic expression of human TAG-1/axonin-1. Eur J Biochem 211: 329–339

Hughes JA, Hurlbert RE, Rupp RA, Spence KD (1983) Bacteria-induced haemolymph proteins of *Manduca sexta* pupae and larvae. J Insect Physiol 29: 625–632

Hultmark D (1993) Immune reactions in *Drosophila* and other insects: a model for innate immunity. Trends Genet 9: 178–183

Hunkapiller T, Hood L (1989) Diversity of the immunoglobulin gene superfamily. Adv Immunol 44: 1–63

Hunkapiller T, Goverman J, Koop BF, Hood L (1989) Implications of the diversity of the immunoglobulin gene superfamily. Cold Spring Harbor Symp Quant Biol 54: 15–29

Hurlbert RE, Karlinsey JE, Spence KD (1985) Differential synthesis of bacteria-induced proteins of *Manduca sexta* larvae and pupae. J Insect Physiol 31: 205–215
Kanost MR, Zepp MK, Ladendorff NE, Andersson LA (1994) Isolation and characterization of a hemocyte aggregation inhibitor from hemolymph of *Manduca sexta* larvae. Arch Insect Biochem Physiol 27: 123–136
Ladendorff NE, Kanost MR (1990) Isolation and characterization of bacteria-induced protein P4 from hemolymph of *Manduca sexta*. Arch Insect Biochem Physiol 15: 33–41
Ladendorff NE, Kanost MR (1991) Bacteria-induced protein P4 (hemolin) from *Manduca sexta*: a member of the immunoglobulin superfamily which can inhibit hemocyte aggregation. Arch Insect Biochem Physiol 18: 285–300
Marchalonis JJ, Schluter SF (1990) Origins of immunoglobulins and immune recognition molecules. BioScience 40: 758–768
Rasmuson T, Boman HG (1979) Insect immunity V. Purification and some properties of immune protein P4 from hemolymph of *Hyalophora cecropia*. Insect Biochem 9: 259–264
Ryan RO, Cole KD, Kawooya JK, Wells MA, Law JH (1988) Identification and characterization of a novel postlarval hemolymph protein from *Manduca sexta*. Arch Insect Biochem Physiol 9: 81–90
Schmidt O, Faye I, Lindström-Dinnetz I, Sun S-C (1993) Specific immune recognition of insect hemolin. Dev Comp Immunol 17: 195–200
Springer TA (1990) Adhesion receptors of the immune system. Nature 346: 425–434
Sun S-C, Faye I (1992) Cecropia immunoresponsive factor, an insect immunoresponsive factor with DNA-binding properties similar to nuclear-factor κB. Eur J Biochem 204: 885–892
Sun SC, Lindström I, Boman HG, Faye I, Schmidt O (1990) Hemolin: an insect-immune protein belonging to the immunoglobulin superfamily. Science 250: 1729–1732
Trenczek T (1988) Injury and immunity in insects. In: Sehnal F, Zabza A, Denlinger DL (eds) Endocrinological frontiers in physiological insect ecology. Wroclaw Tech Univ Press, Wroclaw, pp 369–378
Trenczek T, Faye I (1988) Synthesis of immune proteins in primary cultures of fat body from *Hyalophora cecropia*. Insect Biochem 18: 299–312
Volkmer H, Hassel B, Wolff JM, Frank R, Rathjen FG (1992) Structure of the axonal surface recognition molecule neurofascin and its relationship to a neural subgroup of the immunoglobulin superfamily. J Cell Biol 118: 149–161
Wang Y, Willott E, Kanost MR (1995) Organization and expression of the hemolin gene, a member of the immunoglobulin superfamily in an insect, *Manduca sexta*. Insect Mol Biol 4: 113–123
Williams AF, Barclay AN (1988) The immunoglobulin superfamily — domains for cells surface recognition. Annu Rev Immunol 6: 381–405

Chapter 7

The Interface Between Invertebrates and Vertebrates: Complement vs Ig

R.L. Raison

Contents

1 Introduction

Perhaps the most tantalizing question to be answered in our pursuit of an in-depth understanding of the evolutionary development of the immune system is that of the origins of adaptive humoral immunity. Immunoglobulins, encoded by sets of rearranging genes, are the mediators of adaptive humoral immunity and these molecules, together with the genes that encode them, have been extensively characterized down to and including the phylogenetic level of the elasmobranchs. Failure to conclusively demonstrate adaptive humoral immune function in the primitive vertebrates and invertebrates raises additional questions about the origins of the rearranging Ig gene system and the humoral defense mechanisms employed by the multitude of species that encompass this group of organisms. The studies described in this chapter deal in part with the latter issue in the phylogenetically primitive vertebrate group, the cyclostomes.

The evolutionary development of immune function ranges from rudimentary T cell-like reactivity in the porifera and coelenterates (Hildemann et al. 1977) to a complex, integrated system of cellular and humoral effector functions in the

Immunobiology Unit, Department of cell and Molecular Biology, University of Technology, Sydney, PO Box 123, Broadway, NSW, 2007, Australia

Advances in Comparative and Environmental Physiology, Vol. 23

higher vertebrates. Immunocompetence in the vertebrates is characterized by the ability to respond at the cellular (T cell) or the cellular and humoral (T cell and B cell) level to an antigenic stimulus. All vertebrates above and including bony fishes possess well-defined lymphoid tissue such as thymus, spleen and bone marrow. Although lacking the latter, elasmobranchs possess a thymus and spleen and exhibit both cell-mediated and humoral immune functions. However, clear evidence for the divergence of the T and B lineages is available only from the evolutionary level of the teleosts. While functionally distinct, these lymphocyte classes are otherwise so similar that 98% of their total gene expression is identical (Davis et al. 1982). Furthermore, both cell types are derived from common haematopoietic stem cell progenitors in the bone marrow, from which they differentiate clonally. The antigen receptors expressed on these cells, immunoglobulin (Ig) on B cells and the T cell receptor (TcR) on T cells, exhibit marked similarities at the protein structure level, and in the organization, rearrangement and expression of the genes which encode them (Davis and Bjorkman 1988). These observations suggest that the T and B cell lineages arose during evolution from a common ancestral type, and that Ig and TcR diverged from a common ancestral antigen receptor.

As the "most primitive" living vertebrate, the hagfish occupies a key position in the scheme of evolution of immune function, linking the vertebrates with their invertebrate ancestors, the protochordates. Members of the class Agnatha, order Cyclostomata, hagfish are modern representatives of the earliest evolved vertebrates, the ostracoderms (jawless fishes) and lack defined primary or secondary lymphoid organs. Nevertheless, these phylogenetically primitive vertebrates are capable of mounting rudimentary cellular and humoral immune responses (Raison et al. 1978a). In the hagfish, the mixed lymphocyte reaction (MLR), a measure of cell-mediated immune function, is mediated by a lymphocyte-like cell found in abundance in the open circulatory system (Raison et al. 1987). Morphologically-similar cells have also been implicated in allograft rejection (Hildemann and Thoenes 1967), although in hagfish the rejection of allogeneic skin grafts typically assumes a chronic time course, with a median survival time for first set grafts of 72 days. An accelerated and specific secondary response with median survival of 28 days was indicative of a memory component in this cellular response. Survival times for third-party grafts in animals previously sensitized to second-party grafts were interpreted as indicating the existence of a polymorphic array of transplantation antigens containing many frequently occurring loci (Hildemann 1981).

There is now strong evidence that tunicates mount an adaptive cellular immune response comparable to that seen in the vertebrates. In a study of allogeneic rejection, Raftos has demonstrated alloimmune memory and specificity mediated by a lymphocyte-like cell (Raftos et al. 1988; see also Chap. 4, Vol. 24). The extent of allogeneic incompatibility observed in these studies is indicative of a polymorphic array of histocompatibility antigens in the urochordates.

2 Humoral Immunity in the Agnatha

Humoral immunity to a wide range of antigens has been reported in the cyclostomes by a number of investigators. Hagfish and lamprey have been variously immunized with a range of cellular and soluble antigens and induced activity detected in the serum. However, molecular characterization of the so-called antibodies has yielded a multiplicity of structures. Lamprey "antibody" to bacteriophage f2 was found to be IgM-like, existing in 14S and 6.6S forms (Marchalonis and Edelman 1968). The basic subunit, which consisted of non-covalently associated heavy and light chain-like polypeptides, was estimated as 188 kDa (Marchalonis and Cone 1973). Yet, immunization of lamprey with human erythrocytes yielded an active serum protein of approximately 320 kDa (Pollara et al. 1970). Further characterization revealed a labile molecule consisting of noncovalently associated 75 kDa polypeptides (Litman et al. 1970). Litman proposed that this molecule was in fact related to the agglutinins found in invertebrates and this suggestion was later reenforced by the studies of Hagen et al. who were able to distinguish a naturally occurring haemagglutinin from antibody activity in the serum of *Petromyzon marinus* after immunization with human "O" erythrocytes (Hagen et al. 1985). Induced antigen binding activity was also observed in the sera of hagfish immunized with either KLH (Thoenes and Hildemann 1969) or sheep erythrocytes (Linthicum and Hildemann 1970). However, in neither case was the humoral activity clearly distinguished from the naturally occurring agglutinins of the hagfish.

Subsequent to these studies hagfish were shown to respond to immunization with a streptococcal vaccine (Raison et al. 1978a). Prolonged immunization with group A streptococci revealed induced anti-group A carbohydrate binding activity in approximately 25% of the animals. The binding activity was titratable over a range of concentrations and activities comparable to that displayed by hyperimmune rabbit serum (Fig. 1). While nonimmune hagfish serum displayed

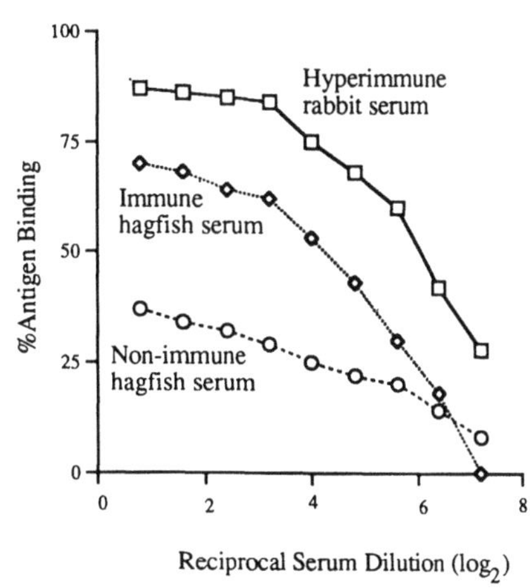

Fig. 1. Induction of humoral antigen binding activity in the hagfish. Hagfish were immunized over a period of several months with a Group A streptococcal whole cell vaccine. Antigen binding activity was detected by radioimmunoassay using ^{125}I-labeled group A carbohydrate. The binding activity in the serum of immunized and nonimmunized hagfish is compared with the activity present in hyperimmune rabbit serum. (Raison et al. 1978a)

significant binding activity to the group A carbohydrate, absorption and sugar inhibition experiments indicated that the anti-streptococcal activity was distinct from the natural haemagglutinin present in hagfish serum. The inducible serum protein, which was found to exhibit specificity for the rhamnose backbone of the group A carbohydrate, was purified and found to have a polypeptide chain structure similar to that óf mammalian IgM. This molecule, referred to as hagfish "antibody", apparently consisted of two polypeptides (~70–75 and ~30 kDa) (Raison et al. 1978b). The subunit dissociated at low concentrations of reducing agent in the presence of denaturing buffer, suggesting that the structure was more labile than that of higher vertebrate immunoglobulin. This was consistent with earlier descriptions of the lamprey "antibody".

3 Structural Characterization of the Hagfish "Antibody"

Using monoclonal and polyclonal antibodies raised against the anti-streptococcal protein purified from hagfish serum we isolated and characterized material affinity purified from nonimmune serum from the Pacific hagfish, *Eptatretus stouti*. Improved analysis revealed the presence of two polypeptides in the heavy chain region migrating with apparent molecular weights of 77 (H1) and 70 kDa (H2) (Hanley et al. 1990) and confirmed a light chain-like component at ~30 kDa. The H chains differed in their amino acid sequences as indicated by the generation of distinct in situ peptide maps. Two-dimensional electrophoretic analysis of the protein under nonreducing and reducing conditions revealed that only a minor proportion of the molecules was covalently linked into the basic subunit of ~200 kDa. Furthermore, a number of potential arrangements of the polypeptide chains were revealed including H1-H1-L, H1-H2-L and H1-H2. Using the isolation and purification techniques described earlier for *E. stouti*, Kobayashi et al. (1985) isolated an antibody-like molecule from the serum of *E. burgeri*. Like the *E. stouti* molecule, this protein exhibited some immunoglobulin-like features but was more labile than conventional immunoglobulin. However, the protein characterized by the Japanese group contained two light chain components and a single heavy polypeptide chain.

The superficial similarity of the hagfish humoral factor with immunoglobulin remained intriguing, although several features, including the apparent lack of significant disulfide bonding, the appearance of multiple heavy and light chain-like components and the possible combinations of these chains in the basic subunit distinguished the hagfish molecule. While amino terminal sequence analysis failed to reveal the heterogeneity normally associated with the variable regions of immunoglobulins (Hanley et al. 1990), analysis of peptides derived from hagfish light and heavy chains yielded sequences that could be aligned over short stretches with human and shark immunoglobulin and human T-cell receptor polypeptide chains (Varner et al. 1991).

Two groups almost simultaneously solved the riddle of the relationship of the hagfish serum protein to higher vertebrate humoral factors. In our laboratory, affinity purified serum protein from *E. stouti* was used to obtain N-terminal and internal amino acid sequence from the larger heavy chain component (Hanley et al. 1992). PCR primers were constructed and yielded a genomic clone containing at least two coding regions. Primers based on this sequence allowed the isolation of a 231 bp PCR fragment from liver RNA. However, comparison of the deduced amino acid sequence with known protein sequences failed to reveal any homology with immunoglobulin or immunoglobulin-like molecules. Unexpectedly, striking similarity was seen when the hagfish protein was compared to the mammalian complement components C3, C4 and C5. Kurosawa and his colleagues at Fujita Health University used "antibody" purified by ion exchange and gel permeation chromatography from the serum of *E. burgeri* to obtain partial amino acid sequence data for the heavy and light polypeptide chains (Ishiguro et al. 1992). Using PCR primers derived from these sequences, they obtained a 2.2 kb fragment from liver mRNA that encoded the three previously described chains of the hagfish molecule. Again, significant sequence similarities were seen with the three complement proteins. Comparison of the nucleotide sequences obtained by the two laboratories left no doubt that they were dealing with the same molecule. Furthermore, amino acid sequences obtained for the putative hagfish "antibody" in three laboratories, and from two species of hagfish (Varner et al. 1991; Hanley et al. 1992; Ishiguro et al. 1992), were encoded in the isolated genes.

Analysis of genomic and cDNA sequence data for the hagfish serum protein indicates that it is compatible with the three polypeptide chain structure previously described. The molecule is encoded by a single mRNA species of approximately 5 kb, the three polypeptides apparently arising from post-translational cleavage of a precursor molecule. Sequence comparisons indicate equivalence of the 77, 70 and 30 kDa hagfish chains with the β, α and γ chains respectively of mammalian C4. Furthermore, the post-translational modification of the hagfish protein includes removal of 77 amino acids encoded at the N-terminus of the α chain. This appears analogous to the similarly sized anaphylotoxins cleaved from the α chains of mammalian C3, C4 and C5 (Fearon and Wong 1983; Larsen and Henson 1983). However, while the proteolytic cleavage of these fragments in mammalian complement components is a consequence of activation of the complement cascade, the cleaved form appears to be the normal serum component in the hagfish. The trigger for this cleavage in the hagfish is not known. While overall sequence comparison shows the greatest similarity of the hagfish protein with mammalian C3, the three chain structure is more similar to that of C4. In addition, mammalian C3 and C4 contain a thioester bond in the α chain. This bond is activated by the cleavage of the C3a and C4a fragments, and plays a key role in the binding of the C3b and C4b fragments to target cells (Fearon and Wong 1983). Similarity of the gene sequences of the α chains of C3 and C4 with the corresponding region of the hagfish gene suggested that this structure would be conserved in the hagfish molecule (Ishiguro et al. 1992). This was confirmed by radiolabeling experiments (Hanley et al. 1992). Based on these data, a structure for the three polypeptide form of the hagfish complement-like protein (CLP) can be constructed (Fig. 2).

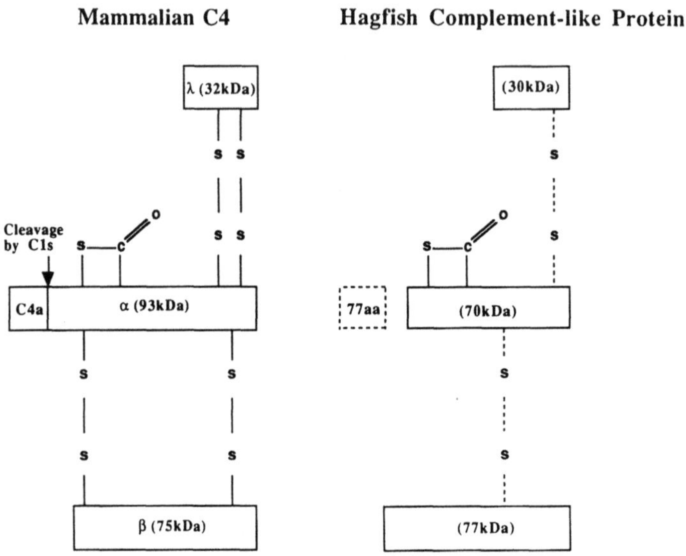

Fig. 2. The proposed structure of hagfish complement-like protein (CLP) is compared to that of mammalian C4. In C4, the α chain is cleaved by activated C1s to yield the 76 amino acid fragment C4a. Comparison of amino acid sequence data for the α chain equivalent of CLP with the cDNA gene sequence indicates that a 77 amino acid fragment (shown in *dotted outline*) is post-translationally removed from the nascent polypeptide chain. The number and positioning of the disulfide bonds in CLP have not been determined

4 Functional Properties of Hagfish Complement-Like Protein

The earlier finding that the hagfish CLP, isolated from immunized animals, exhibited binding affinity for carbohydrate groups in the streptococcal cell wall implied a humoral defense role for this molecule. Furthermore, given the relationship of this molecule to complement components involved in enhancement of immune function via opsonization, we examined the potential for CLP to mediate this function. Affinity purified CLP from nonimmune animals was shown to bind to the surface of bacteria (streptococci) and yeast and enhance the phagocytosis of these cells by hagfish monocytes (Hanley et al. 1992). The binding and opsonizing properties of CLP were inhibited by coincubation with rhamnose or mannose, but not by a range of other monosaccharides (Raftos et al. 1992). The specificity for rhamnose was of particular interest as it was consistent with our earlier findings with respect to the protein identified in, and purified from, the serum of hagfish immunized with streptococci. It is thus evident that CLP exists at significant constitutive levels in the serum of hagfish and that its production, probably from cells in the liver, can be increased in response to antigen challenge.

The binding and opsonizing properties of hagfish CLP are similar to those displayed by the thioester-containing mammalian complement components C3 and C4 (Fearon and Wong 1983), as well as the C4-like molecule identified in the lamprey, *Lampreta japonica* (Nonaka et al. 1984). However, some aspects of the hagfish protein distinguish it from the mammalian counterpart. Mammalian C3 and C4 must undergo proteolytic cleavage to activate the thioester group which then covalently interacts with hydroxyl or amino determinants on the target cell surface. The activity of hagfish CLP appears to require no prior activation as the material isolated from normal serum is active, even in the absence of additional serum proteins. In addition, while the activated mammalian complement component may covalently attach to carbohydrates via the thioester group (Law and Levine 1981), the reactivity does not appear to exhibit the restricted sugar specificity seen with CLP. Indeed, the binding of CLP to carbohydrates is lectin-like in that it is divalent cation-dependent (Raftos et al. 1992). Thus, a direct link between the presence of the thioester bond on the α-like chain, and the carbohydrate binding specificity of CLP cannot be made at this time.

CLP mediated opsonization occurs via a receptor on the surface of hagfish monocytes (Raison et al. 1994). A monoclonal antibody, raised against monocytes present in the hagfish peripheral blood leukocyte population, inhibited the phagocytosis of CLP coated target cells. In the reciprocal experiment, zymosan activated CLP blocked the interaction of the monoclonal antibody with monocytes. The monoclonal antibody immunoprecipitated a single polypeptide chain of approximately 100 kDa from the surface of radiolabeled leukocytes. This molecule migrated with an apparent higher molecular weight when electrophoresed under reducing and denaturing conditions, suggesting that it contains significant internal disulfide bonds. In humans the receptor which mediates the enhanced phagocytosis of C3 coated target cells is CR3, a member of the integrin family of adhesion molecules. Members of this family have been described in invertebrates (Bogaert et al. 1987) and it is tempting to speculate that the opsonic receptor identified on hagfish monocytes may belong to this phylogenetically primitive group of proteins. However, CR3 consists of two subunits and thus, at least superficially, does not appear related to the hagfish receptor. Two other forms of C3 receptor have been identified in mammals; CR1, the ligand for which is C3b, and CR2 which is able to bind the degradation fagments of C3b (Fearon and Wong 1983). Again, the molecular characteristics of these receptors do not immediately suggest a phylogenetic linkage to the hagfish receptor.

5 Complement in the Agnatha

While there is circumstantial, functional evidence for the existence of a rudimentary alternative pathway of complement activity in the invertebrates, no molecules exhibiting structural homology with known complement components have been isolated and characterized in these species (Farries and Atkinson 1991). Thus, the first report of a phylogenetically primitive complement component was that of a C3-like protein from the lamprey (Nonaka et al. 1984). A similar molecule was isolated from the serum of the hagfish and shown to exhibit both primary and secondary structural features in common with human C3 (Fujii et al. 1992). The hagfish C3 consisted of two disulfide-linked polypeptide chains of 115 and 77 kDa. Amino-terminal sequence analysis of the 77-kDa chain of hagfish C3 revealed identity with the 77-kDa chain from the previously described CLP molecule. More recently, Fujii et al. (1993) reported a variant form of the hagfish C3 molecule with a three polypeptide chain structure identical to that of CLP. The variant form of C3 was found to arise during purification from hagfish serum presumably by proteolytic cleavage of the C3 α chain to 77 and 30 kDa polypeptides. However, it appears that only the variant form of hagfish C3 undergoes this form of cleavage to produce the three-chain molecule. Taken together, these findings suggest that complement proteins bearing primary and secondary structural features in common with mammalian proteins C3 and C4 are produced in the hagfish. Furthermore, the available sequence data strongly indicate that the different molecular forms of hagfish complement identified to date may arise through post-translational processing of the product of a single mRNA species.

The C3- and C4-like (CLP) proteins of lamprey and hagfish, like the human C3 component, bind to yeast cells and act as a ligand for opsonization by phagocytic cells (Nonaka et al. 1984; Fujii et al. 1993; Raison et al. 1994). Lytic function associated with complement components has not been demonstrated in the cyclostomes, suggesting that they lack the terminal pathway proteins, C5–9 (Farries and Atkinson 1991).

6 Chemotaxis in the Hagfish

Chemotaxis, the directed migration and recruitment of effector cells, is a fundamental component of the immune and inflammatory responses. A number of chemoattractants have been identified in mammals and include IL-8, C5a and the bacterial tripeptide, formyl methionyl leucyl phenylalanine (fMLP) (Hugli 1989). In particular, chemotaxis of phagocytic cells plays a major role in defense against infection.

The phylogenetic conservation of chemotactic function has been observed through the ability of mammalian C5a to enhance the migration of leukocytes in the shark (Obenauf and Hyder-Smith 1985). Recently, responsiveness of hagfish leukocytes to endogenous and exogenous chemoattractants has been demonstrated (Newton et al. 1994). Hagfish monocytes and granular leukocytes migrated in a concentration gradient of both human C5a and LPS-activated hagfish plasma. Thus, activation of hagfish plasma generated a chemotactic product which reacted in a manner analogous to human C5a. In mammals, the 74 amino acid anaphylotoxin C5a, is cleaved from the amino terminus of the α chain of C5. The corresponding chain of hagfish CLP exhibits 32% sequence identity and 43% sequence similarity (i.e., allowing for conservative substitutions) with human C5 (Hanley et al. 1992; Ishiguro et al. 1992). It is therefore possible that LPS activation results in the generation of a C5a-like fragment from the hagfish α chain homologue. This fragment appears to be the endogenous ligand for a receptor on hagfish leukocytes which also reacts with human C5a. This speculation is supported by the obvious evolutionary relatedness of mammalian C3, C4 and C5 and the primary structural similarity of CLP with these three components.

7 Conclusion

It is now clear that the hagfish serum protein previously described as "antibody" or "immunoglobulin" is in fact a complement related protein, or series of proteins, bearing strong primary and secondary structural similarities to the mammalian C3, C4 and C5 complement proteins. Given the identification of a similar complement protein in lamprey it would appear reasonable to conclude that this situation extends through the cyclostomata. Thus, humoral immunity in these primitive vertebrates is mediated, at least in part, by complement-like proteins that function as broadly reacting opsonins.

Considerable effort has been directed toward the question of the presence of immunoglobulin in the cyclostomes. Despite numerous reports of antibody-like activity in serum after immunization with cellular and soluble antigens, a molecule with definitive structural features of immunoglobulin has not been isolated. A number of techniques have been used in attempts to isolate immunoglobulin genes from cyclostomes. Although the use of mammalian immunoglobulin variable region probes to isolate homologous genes in other species has been highly successful down to, and including, the cartilaginous fishes (Hinds and Litman 1986), this approach has failed to identify immunoglobulin-like genes in the hagfish. Similarly, the use of phylogenetically conserved blocks of sequence in a PCR strategy has not yielded immunoglobulin-like genes from either lamprey or hagfish (Ishiguro et al. 1992) despite the success of the approach in isolating MHC genes from carp (Hashimoto et al. 1990) and shark (Hashimoto et al. 1992). When viewed overall, there is at this time no molecular evidence to support the existence of immunoglobulin-like genes in the most primitive vertebrates. However, given the identification of immunoglobulin-like domain structures in molecules such as amalgam, fasciclin II and haemolin which are found in insects (Harrelson and Goodman 1988; Seeger et al. 1988; Sun et al. 1990), it would appear inevitable that members of the immunoglobulin gene superfamily will ultimately be identified in the cyclostomes. This speculation notwithstanding, there is currently no definitive evidence for an adaptive humoral immune response based on rearranging immunoglobulin-like genes at this level of phylogeny. While the data do suggest a T cell-like role for a population of circulating leukocytes in the hagfish, it appears that humoral immunity in cyclostomes is a function of innate, nonadaptive mechanisms.

References

Bogaert T, Brown N, Wilcox M (1987) The Drosophila PS2 antigen is an invertebrate integrin that, like the fibronectin receptor, becomes localised to muscle attachments. Cell 51: 929–940
Davis MM, Bjorkman P (1988) T cell antigen receptor genes and T cell recognition. Nature 334: 395–402
Davis MM, Cohen DI, Nielsen EA, DeFranco AL, Paul WE (1982) The isolation of B and T cell-specific genes. In: Vitetta ES, Fox CF (eds) B and T cell tumours. Academic Press, New York, pp 215–220
Farries TC, Atkinson JP (1991) Evolution of the complement system. Immunol Today 12: 295–300
Fearon DT, Wong WW (1983) Complement ligand-receptor interactions that mediate biological responses. Annu Rev Immunol 1: 243–271
Fujii T, Nakamura T, Sekizawa A, Tomonaga S (1992) Isolation and characterisation of a protein from hagfish serum that is homologous to the third component of the mammalian complement system. J Immunol 148: 117–123
Fujii T, Nakamura T, Tomonaga S (1993) Identification and characterisation of a variant of the third component of complement (C3) in hagfish serum. Zool Sci (Suppl) 10: 83
Hagen M, Filosa MF, Youson JH (1985) The immune response in adult sea lamprey, *Petromyzon marinus*: the effect of temperature. Comp Biochem Physiol 82A: 207–210
Hanley PJ, Seppelt IM, Gooley AA, Hook JW, Raison RL (1990) Distinctive Ig H chains in a primitive vertebrate, *Eptatretus stouti*. J Immunol 145: 3823–3828
Hanley PJ, Hook JW, Raftos DA, Gooley AA, Trent R and Raison RL (1992) Hagfish humoral defense protein exhibits structural and functional homology with mammalian complement proteins. Proc Natl Acad Sci USA 89; 7910–7914
Harrelson AL, Goodman CS (1988) Growth cone guidance in insects: fasciclin II is a member of the immunoglobulin superfamily. Science 242: 700–703
Hashimoto K, Nakanishi T, Kurosawa Y (1990) Isolation of carp genes encoding major histocompatibility complex antigens. Proc Natl Acad Sci USA 87: 6863–6867
Hashimoto K, Nakanishi T, Kurosawa Y (1992) Identification of a shark sequence resembling the major histocompatibility complex class I $\alpha 3$ domain. Proc Natl Acad Sci USA 89: 2209–2212
Hildemann WH (1981) Immunophylogeny: from sponges to hagfish to mice. In: Hildemann WH (ed) Frontiers in immunogenetics. Elsevier, New York, pp 3–19
Hildemann WH, Thoenes GH (1967) Immunological responses of Pacific hagfish. I. Skin transplantation immunity. Transplantation 7: 506–521
Hildemann WH, Raison RL, Chenung G, Hull CJ, Akaka L and Okamoto J (1977) Immunological specificity and memory in a scleractinian coral. Nature 270: 219–223
Hinds KR, Litman GW (1986) Major reorganisation of immunoglobulin V_H segmental elements during vertebrate evolution. Nature 320: 546–549
Hugli, TE (1989) Chemotaxis. Curr Opin Immunol 2: 19–27
Ishiguro H, Kobayashi K, Suzuki M, Titani K, Tomonaga S, Kurosawa Y (1992) Isolation of a hagfish gene that encodes a complement component. EMBO J 11: 829–837
Kobayashi KS, Tomonaga S, Hagiwara K (1985) Isolation and characterisation of immunoglobulin of hagfish, *Eptatretus burgeri*, a primitive vertebrate. Mol Immunol 22: 1091–1097
Larsen GL, Henson PM (1983) Mediators of inflammation. Annu Rev Immunol 1: 335–360
Law SK, Levine RP (1981) Binding reaction between the third human complement protein and small molecules. Biochemistry 20: 7457–7463
Linthicum DS, Hildemann WH (1970) Immunologic responses of Pacific hagfish. III. Serum antibodies to cellular antigens. J Immunol 105: 912–918
Litman GW, Frommel D, Finstad J, Howell J, Pollara BW, Good RA (1970) The evolution of the immune response. VIII. Structural studies of the lamprey immunoglobulin. J Immunol 105: 1278–1285
Marchalonis JJ, Cone RE (1973) The phylogenetic emergence of vertebrate immunity. Aust J Exp Biol Med Sci 51: 461–488

Subject Index*

**Acknowledgment*. Expressions of gratitude are extended to Andrew J. Yang, volunteer student research program for his invaluable help in preparing the index.

MIX
Papier aus verantwortungsvollen Quellen
Paper from responsible sources
FSC® C105338

FSC
www.fsc.org

If you have any concerns about our products,
you can contact us on
ProductSafety@springernature.com

In case Publisher is established outside the EU,
the EU authorized representative is:
Springer Nature Customer Service Center GmbH
Europaplatz 3, 69115 Heidelberg, Germany

Printed by Libri Plureos GmbH
in Hamburg, Germany